Dragonflies

Steve Brooks

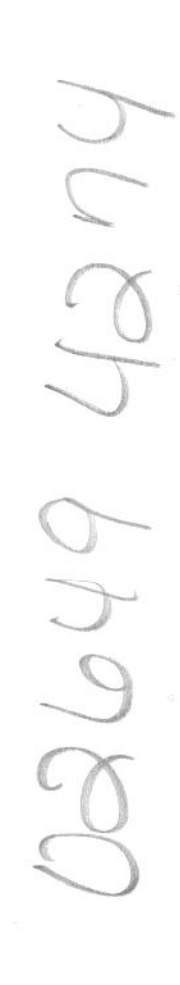

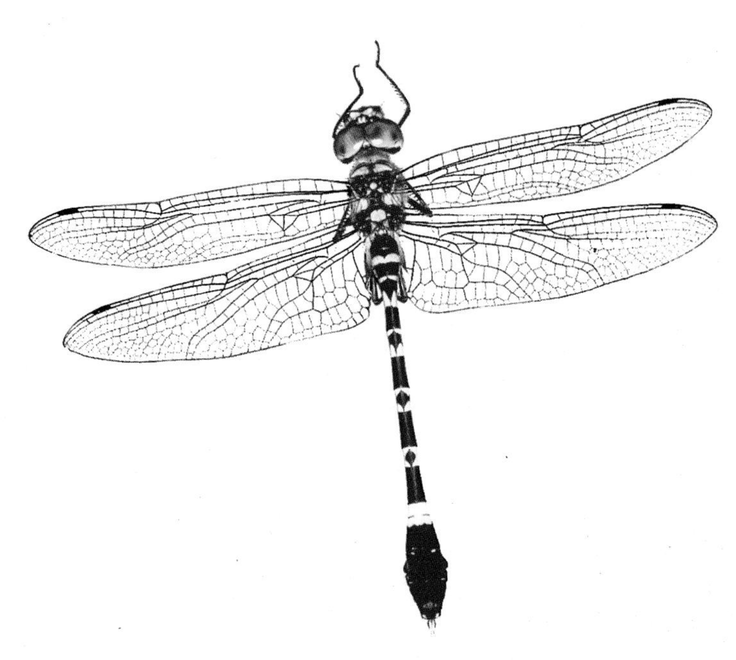

SMITHSONIAN BOOKS, WASHINGTON, D.C.
IN ASSOCIATION WITH THE NATURAL HISTORY MUSEUM, LONDON

Published in the United States of America
by Smithsonian Books in association with
The Natural History Museum, London
Cromwell Road
London SW7 5BD
United Kingdom

Library of Congress Cataloging-in-Publication Data
Brooks, S. J.
Dragonflies / Steve Brooks.
p. cm.
Includes bibliographical references and index.
ISBN 1-58834-064-3 (pbk. : alk. paper)
1. Dragonflies. I. Title.
QL520.B76 2003
595.7'33—dc21 2002036456

Manufactured in Singapore, not at government expense
10 09 08 07 06 05 04 03 5 4 3 2 1

Edited by Jonathan Elphick
Designed by Mercer Design
Reproduction and printing by Craft Print, Singapore

Front cover main image:
Southern hawker *Aeshna cyanea*
Front cover inset: Beautiful demoiselle *Calopteryx virgo*
On the back cover and title page:
A river emerald *Macromia picta*

Contents

Preface

Most of us are aware of dragonflies and damselflies. Their aerial agility and vivid colours are evocative of the long, hot days of summer, of shimmering lakes and dappled shade by trickling streams. But fewer people will have looked beyond the flash of blue, red, green or yellow to consider what each dragonfly is doing.

This book takes you into the world of these fascinating insects and introduces you to their complex lifestyles – from the ferocious larval stages, lurking amongst plants and debris in lakes and rivers, to the breathtaking adults swooping after prey or fighting rivals to defend their territories. Dragonflies and damselflies have lived on earth for hundreds of millions of years, and there are thousands of different species living in watery environments throughout the world. Gaining a better understanding of the needs of dragonflies helps to deepen one's appreciation of the value of wetlands, from the smallest farm pond to the largest marsh, lake or river.

The task of gathering information for the writing of this book has been made far easier by the recent publication of two other books on dragonflies. The first, by Philip Corbet, comprehensively reviews their behaviour and ecology, while the other, by Jill Silsby, takes the reader on a safari of the major groups of dragonflies of the world. For readers who wish to pursue their interest in dragonflies further, both these books are strongly recommended. (See Further Information, page 96)

Author

Steve Brooks's respect for dragonflies began as a boy, when the first dragonfly larva he caught in a local pond ate every other creature he had collected in his jam-jar.
He has been fortunate to follow his boyhood interest in dragonflies and wetland natural history into a career at The Natural History Museum, London, where he has worked since 1979. Steve is author of the *Field guide to the dragonflies and damselflies of Great Britain and Ireland*, and has written many scientific articles on dragonflies. He is a member of the British Dragonfly Society (BDS), and has been editor of the BDS Journal and a member of the BDS Conservation Group. His research into dragonflies has taken him to much of Europe, southern and western Africa, and Central America.

What are dragonflies and damselflies?

Dragonflies and damselflies belong to an order (major group) of insects known by the scientific name of Odonata. This is derived from the Greek word *odon*, meaning 'tooth', and refers to the strongly toothed jaws that all odonate species possess. Systematists (the scientists who study the evolutionary relationships between groups of animals and plants) recognize three suborders of the living species of Odonata. The Anisoptera (comprising all the dragonflies), the ancient suborder Anisozygoptera and the Zygoptera (comprising all the damselflies). Collectively, they are referred to as the Odonata, or odonates.

About 5500 species of odonates are known throughout the world. Some 2500 of these are damselflies, divided among 20 families. The remaining 2700 species are dragonflies, included in eight families. The suborder Anisozygoptera includes only two living species, both in the same family,

LEFT **The keeled skimmer *Orthetrum coerulescens* is a typical dragonfly with large eyes, hindwings broader than the forewings, and rests with wings outspread.**

with one in the Himalayan Mountains and the second in Japan, although the fossil record shows that this group was much more diverse in the past.

Distribution and habitat

Odonates occur throughout most of the world, but are absent from Antarctica and some Arctic islands. There are no resident species at all in Iceland, although occasionally migrants from North Africa are recorded there. Odonates are most abundant and diverse in tropical regions. For example, the small tropical Central American country of Costa Rica, (50,000 km^2/19,300 sq miles in area) supports at least 250 species. By comparison, only just over half that number (140 species) are known to occur in subtropical Florida (area 140,000 km^2/ 54,000 sq miles), and just 34 species in England (area 130,000 km^2/50,200 sq miles). Odonata are usually seen near lakes, ponds, rivers, streams and other water bodies, where most species lay their eggs and their larvae develop. Standing-water habitats support the highest abundance of Odonata, but the greatest diversity of species is associated with upland tropical streams. Odonata breed in a wide variety of aquatic environments – most commonly in permanent lakes and ponds, rivers and streams. But some species need specialist habitats that include forest leaf-litter, water-filled leaf axils or rot-holes in trees, the faces of waterfalls, marine rock pools and salt marshes, temporary water bodies and shallow seepages or trickles.

Distinctive features

Adult odonates can be distinguished from other insects by their long, ten-segmented

Living in two worlds

Odonata are exopterygotes (the word refers to the presence of wing sheaths visible externally in later larval stages). Like all exopterygotes, odonates do not have a pupal stage in the life cycle. In many exopterygote insects, including bugs, aphids, grasshoppers and cockroaches, the larval stage lives in the same habitat as the adult and is broadly similar in appearance. But Odonata, together with their close relatives the mayflies (Order Ephemeroptera), occupy completely different habitats as larvae and adults (water and land), and so avoid the potential of competing for the same resources. In endopterygote insects, such as beetles, flies, butterflies and moths, the wing sheaths are never visible externally in the larval stage, and each life stage is strikingly different in appearance and behaviour. In contrast to the larval stages of odonates and mayflies, the larva, caterpillar or maggot looks nothing like the adult stage. Like them, however, it occupies a completely separate habitat from that of the adult. When the larval stage of endopterygote insects has completed development it passes into the chrysalis or pupal stage. This stage is inactive and does not feed, but within the pupa the larval structure is broken down and reconstituted into the adult insect, which emerges, fully formed, from the pupa after several weeks or months with a completely different body-form.

abdomens, their two pairs of large wings with a dense network of fine veins, and their large eyes and short antennae. Most adults are brightly coloured. They are active during the day, especially on hot summer days, and many species are fast and agile in flight. All odonates are fierce predators, both as adults and larvae.

Adult dragonflies are characterized by the forewings being narrower than the hindwings: the name Anisoptera is derived from the Greek words meaning 'unequal wings'. When at rest, dragonflies perch with their wings held stiffly open. They have a powerful darting or swooping flight. In addition, a dragonfly's head is almost fully occupied by a pair of enormous compound eyes, which give it a globular appearance.

Adult damselflies, on the other hand, have forewings and hindwings of a similar shape: the name Zygoptera is derived from the Greek words meaning 'equal wings'. Most species of damselfly perch with their wings held closed over the top of the abdomen, although some hold the wings partially open, or even fully open like dragonflies. Unlike that of dragonflies, the flight of damselflies is weak and fluttering. Also, their eyes are smaller, and are situated on the side of the rather rectangular head.

RIGHT **The banded demoiselle *Calopteryx splendens* is a typical damselfly with small eyes, forewings and hindwings of similar shape, and rests with wings closed.**

FAR LEFT **In this hawker dragonfly larva (Family Aeshnidae) you can see the five stout spines at the tip of the abdomen that are typical of dragonfly larvae.**

LEFT **The three leaf-like appendages at the tip of the abdomen are a characteristic feature of damselfly larvae, as in this red-eyed damselfly *Erythromma najas*.**

The larvae of most odonate species occur in freshwaters, where they can be common among submerged plants or among the debris and silt at the bottom. Damselfly larvae are slender insects with three leaf-like structures at the tip of the abdomen, whereas dragonfly larvae are more robust and have a group of five short spines at the end of the abdomen.

Odonate larvae may take up to eight years to complete development. Compared with the larval stage, the adults of most odonate species are relatively short-lived, and often survive for only a few weeks or months. During this time they must disperse, eat, find mates and produce offspring. The biology and ecology of larval and adult Odonata are complex (see chapters following).

The fossil record

Odonates were among the first winged insects to evolve. About 325 million years, ago giant Protodonata flew in the Upper Carboniferous forests. These huge insects, with a wingspan of up to 1 m (3 1/4 ft), are the largest insects ever known. They are thought to be the ancestors of the Odonata. Their massive wings had a dense network of veins, which superficially resembled the wings of odonates. However, the protodonates lacked the distinctive node, or notch, in the middle of the leading edge of each wing. They were also without a pterostigma, the thickened and pigmented lozenge, made of the tough protein chitin (see page 33), that is situated on the leading edge near the tip of each wing in

Odonata. These two structures are unique to Odonata, and their absence in Protodonata indicates that these enormous insects were only distantly related to Odonata.

Little is known about the appearance of the head and bodies of adult Protodonata. This is because in most fossils, only the wings are preserved, so scientists can only speculate about their lifestyle. However, the eyes were small and the antennae long in the protodonate *Namurotypus sippeli*, and the legs do not bear the long spines typical of Odonata, which suggests that this species was not a proficient aerial predator. Similarly, the larval stages of Protodonata are unknown, so researchers do not know whether they lived on land or in water. It is known, however, that insects evolved on land, so this implies that the ancestors of Odonata did have terrestrial larvae and that the aquatic habit of modern odonate larvae evolved later.

Disappearance of the Protodonata

About 245 million years ago, as one of the many casualties of the great extinction event at the end of the Permian period, the Protodonata disappeared completely from the fossil record. Nobody is sure why, just as no one can be certain what allowed these insects to grow so large.

The answer to these puzzles may be related to oxygen levels in the atmosphere. During the Carboniferous period (350–300 million years ago), the proportion of oxygen in the air was at an all-time high. Since insects must rely on diffusion to transport oxygen around their bodies, high atmospheric oxygen concentrations could have allowed insects larger than those present today to evolve. Oxygen levels declined by 20% after the Carboniferous, and this may have led to the demise of all giant insects, including the Protodonata. Alternatively, the relatively thin air of the Carboniferous may have favoured the evolution of insects with a large wing area. Or perhaps these large ungainly insects became easy prey for reptiles that began to evolve rapidly at that time.

ABOVE **This fossil of the Protodonatan *Whalleyala bolsoveri* was found in the coal seams of Yorkshire, UK. The species lived during the Carboniferous and had a wing length of about 90 mm (3½ in).**

The rise of the Odonata

The first Odonata appear in the fossil record during the Lower Permian period, in rocks thought to be about 250 million years old. These insects are remarkably similar in appearance to modern Odonata – a clear demonstration of how well suited the adult body form of Odonata is for the lifestyle of aerial predator. The first examples of adult and larval Odonata belonging to families still represented today appeared about 200 million

LEFT **This fossil of *Cymatophlebia longialata* was discovered in Jurassic limestone in Solenhofen, Germany. The presence of a pterostigma at the tip of each wing, and a nodus in the middle of the leading edge of each wing, indicates that it is a true odonatan.**

years ago, during the Jurassic period. The fossil record shows that Odonata were once far more diverse than they are now, and several other suborders in addition to the three known today occurred during the time of the dinosaurs, between 250 and 65 million years ago. These are all now extinct.

Body form

In common with all insects, the Odonata have bodies that are divided into three main parts. These are the head, thorax and abdomen. The divisions can be clearly seen if you look at one of the photographs of the odonates in this book.

Head

The front of the head includes the frons, and clypeus. Behind the frons is a pair of short antennae, and between these are three simple eyes (the ocelli), arranged in a triangle. The most obvious structures on the head are the huge compound eyes, which are made up of thousands of individual lenses (the ommatidia). Below and in front of the eyes are the mouthparts comprising the labrum (an 'upper lip') and the labium (a 'lower lip'). Between these structures are a pair of toothed mandibles. These hinge sideways and shear across each other, acting like a pair of serrated scissors to impale and slice up prey.

Thorax

The thorax is connected to the head by a thin 'neck', which allows the head to swivel with great flexibility. The thorax comprises three segments: the first is the prothorax, to which is attached the first pair of legs. The second and third segments, the mesothorax and metathorax, are fused in Odonata, and together are termed the synthorax. To these segments are attached the middle and hind legs, and also the forewings and hindwings. The synthorax is massively swollen, to enclose the large and powerful wing muscles. The synthorax is uniquely twisted in adult odonates, in such a way that the upper side is inclined backwards. The effect of this is to push the legs forwards. The legs of adult Odonata are equipped with rows of long, stout spines and end in a pair of strong claws.

Abdomen

The abdomen comprises ten segments, each one being about two to three times as long as it is broad. On the underside of the second abdominal segment in male odonates are the secondary sex organs, comprising paired lobes and hooks (the hamules), together with the penis. At the base of segment eight in males is the genital opening, and in females the ovipositor.

At the apex of the abdomen are the anal appendages. These are largely unmodified in adult females, but are hooked in adult males. Larval damselflies bear three flattened, leaf-like appendages at the tip of the abdomen, but larval dragonflies are equipped with five short spines instead.

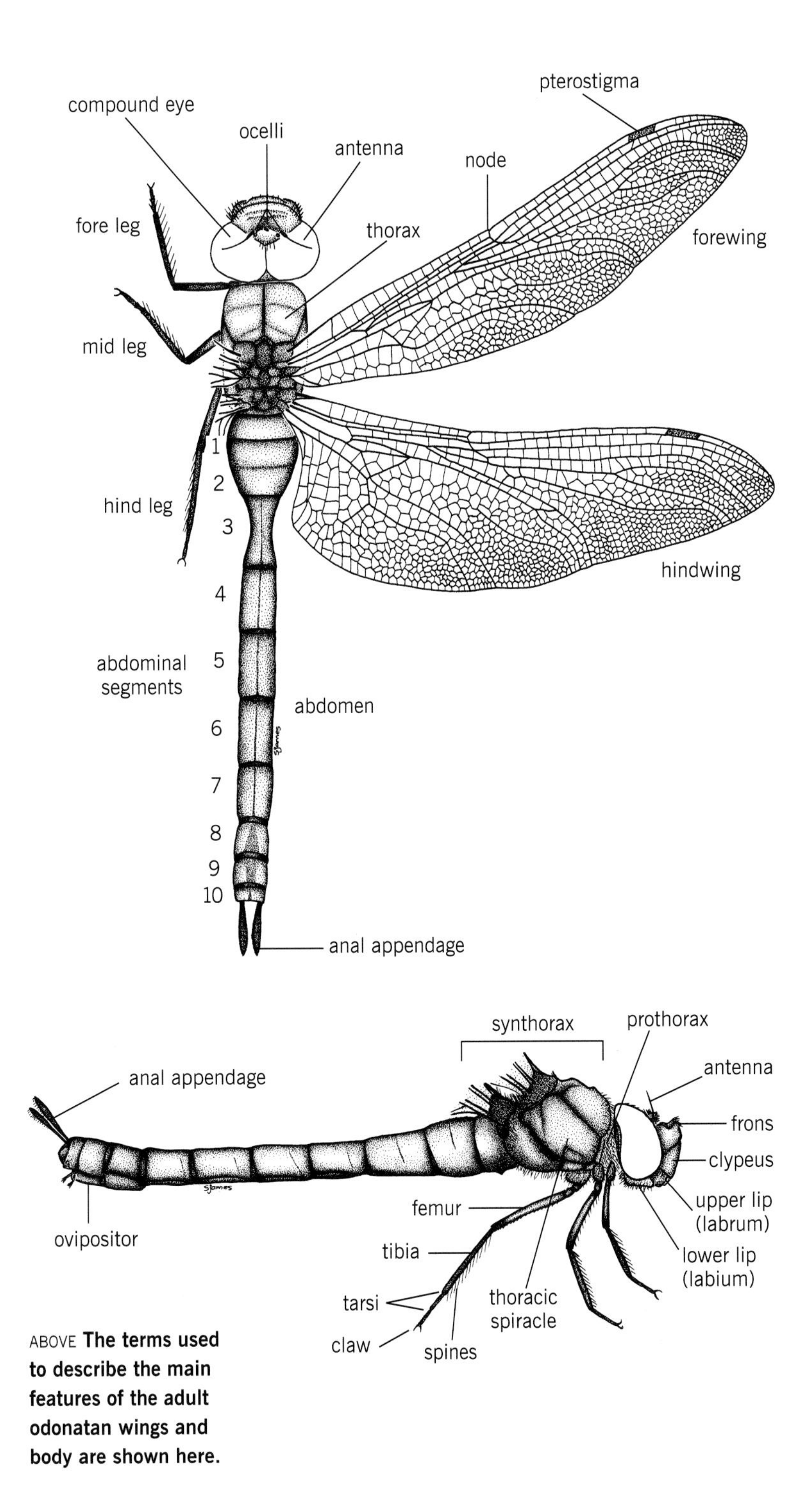

ABOVE **The terms used to describe the main features of the adult odonatan wings and body are shown here.**

The scientific classification of Odonata

The study of systematics is an attempt by biologists to bring order to the natural world. It is a means of imposing a hierarchy on living plants and animals, with its basis in evolution, by grouping together those organisms that are thought to be most closely related to each other. Species within these groups are usually similar in appearance, and can be expected to have similarities in ecology.

A name is given to each level in the hierarchy. A species represents a group of individuals forming a single population (which may be spread over a large geographical area) that is capable of interbreeding and producing viable offspring. Each species is given a unique combination of two names, known as its binomial. The Swedish naturalist Carl von Linné (Linnaeus) first proposed this system of classification, in 1758.

The highest level in this classification is the kingdom, of which there are five, including plants and animals. The animal kingdom is divided into many major groups, called phyla (singular: phylum). Odonata belong to the phylum Arthropoda, which includes all animals with a hard exoskeleton and jointed limbs. The Arthropoda is divided into a number of classes, of which one is the Insecta (insects). The next level in the hierarchy is termed an order. Odonata is one of nearly 30 orders of insects. Others include beetles (Coleoptera), wasps (Hymenoptera), flies (Diptera) and butterflies and moths (Lepidoptera). The order Odonata is split into three suborders: Anisoptera (dragonflies), Zygoptera (damselflies) and Anisozygoptera.

Below the level of the order is the family. All family names end in the suffix *–idae*, for example the dragonfly family Libellulidae (popularly known in Britain as chaser, skimmer and darter dragonflies). Families are themselves divided into subfamilies (with *–inae* endings) and tribes (which are given *–ini* endings).

However, at present, understanding of the systematics of many of the families of Odonata at the subfamily and tribal level is poor, especially within the damselflies. The currently used subfamily and tribal groupings do not withstand rigorous analysis, and do not appear to reflect the probable evolutionary relationships of the Odonata.

Below these higher taxonomic groupings are genera (singular: genus), which comprise groups of closely related species. By convention, genus (or generic) names are written in italics and start with a capital letter, for example *Libellula*, the genus that includes all the chaser dragonflies. Species (or specific) names, such as *depressa*, are also italicized, but these

BELOW **The broad-bodied chaser *Libelulla depressa* is common throughout most of Europe, except northern Britain, Ireland and Scandinavia, and extends into western Asia. This individual has only recently emerged.**

names always begin with a lower-case letter. The species name always appears with the genus name, for example *Libellula depressa*, the scientific name of the dragonfly commonly known in Britain as the broad-bodied chaser. The genus name may be abbreviated to its initial letter, in this case as *L. depressa*.

The name of the person who originally described the species and the date of this description are often included after the species name: for example, *Libellula depressa* Linnaeus, 1758. If the species is no longer assigned to the genus in which it was originally described, then the author's name appears in brackets. The full name of a species – the species name in combination with its genus name – is unique.

Species are themselves sometimes further subdivided, into subspecies. This can be helpful if a species shows distinct regional differences in morphology (form and structure) within its geographic range. Each subspecies is given a name, which is added after the species name; for instance, *Libellula fulva fulva*, which occurs over much of western Europe, or *Libellula fulva pontica*, found only in western Turkey. As with specific names, these subspecific names are often abbreviated – in the examples just given, as *L. f. fulva* and *L. f. pontica*. Subspecies are capable of interbreeding and producing intermediate forms.

The specimen upon which the original description of a species was based is called a holotype (or type specimen). Holotypes are important because they serve as the basis for naming and describing new species. These unique specimens are usually deposited in large national museums, such as The Natural History Museum in London, UK, or the Smithsonian Institute in Washington, USA, and are among the most valuable specimens in the museum's collections. The Natural History Museum has about 1200 holotypes of Odonata in its collections, more than any other museum in the world.

Immature stages

People often ask 'How long do dragonflies live?' and are surprised at the brevity of the adult stage, which in many species may be just a few weeks, compared with the egg and larval stages which may last several years. People also often remark on how ugly the larvae seem when the adults are so attractive. In fact Odonata larvae are highly accomplished and ferocious underwater hunters, often the top predators in fishless waters, that can exploit almost every variety of aquatic habitat throughout the world.

Eggs

Odonata protect their eggs from drought, extremes of temperature and predators by adopting one or other of two egg-laying strategies. In the first, known as endophytic oviposition, the eggs are laid individually inside the stems or leaves of living or dead plants. In the second, called exophytic oviposition, they are laid individually, or in a mass (sometimes containing thousands of eggs) surrounded by a protective layer of jelly, directly into water. The jelly is sticky and helps to secure the eggs among submerged plants over which they are laid. This prevents the eggs ending up in unsuitable conditions, such as would result if they were to sink into deep, oxygen-poor or cool, shaded areas of water, or were swept downstream.

Endophytically-laid eggs are narrow and elongate, whereas those deposited exophytically are more rounded. Within the entire order Odonata, eggs range in size from about 480 x 230 µm (0.48 x 0.23 mm/less than 1/50–1/100 in) in one of the smallest dragonflies, *Nannophya pygmaea*, to 770 x 600 µm (0.77–0.60 mm/about 1/30–1/40 in) in one of the largest dragonflies, *Anotogaster sieboldii*.

Hatching rates

The rate at which the eggs hatch depends on the surrounding temperature. The embryos of species living in small shallow pools in tropical countries develop the most quickly,

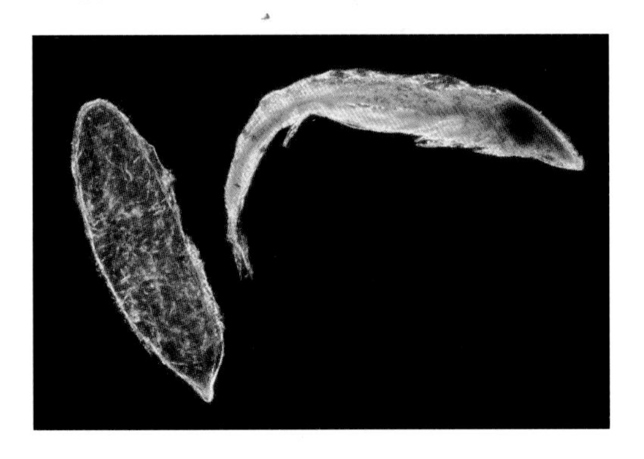

FAR LEFT **This spindle-shaped egg of the emperor dragonfly *Anax imperator* is typical of a species that lays its eggs inside the stems or leaves of plants. Next to the egg is the prolarva.**

LEFT **In species that lay exophytically the eggs are more rounded, like these ones of the common darter *Sympetrum striolatum*.**

and may hatch within just five days of the eggs being laid. In temperate regions, by contrast, 60 days may pass before the egg hatches. In temperate species that lay their eggs towards the end of summer, hatching is often delayed until the following spring. The eggs enter a period of arrested development (known as diapause), in anticipation of low temperatures and reduced food availability during the winter, which would be unfavourable to newly hatched larvae. In these circumstances the egg stage may last 80–230 days. Generally, the embryo will not begin to develop at temperatures below 10°C (50°F) or above 32°C (90°F) with an optimum range of about 20–30°C (68–86°F).

Egg survival

The survival of the eggs may be influenced by several factors. Under normal circumstances, over 90% of eggs laid by odonates are fertile. However, the thickness of the plant tissue into which endophytic eggs are inserted, or the thickness of the jelly in which exophytic eggs are laid, may influence hatching success. Embryos may also be killed by prolonged periods of drought, or by low or high temperatures. Exophytic eggs are particularly vulnerable to these causes of mortality. Tiny parasitic wasps may also attack eggs. In some cases, 90% of eggs may harbour a developing calcidiod wasp larva, although endophytic eggs laid within the thickest part of the plant tissue tend to be less heavily parasitized than those deposited in more exposed places. Odonate eggs, especially those laid exophytically, are susceptible to predation by water mites and fish.

Hatching

Hatching occurs after the fully developed embryo swallows water, drawn through minute holes in the membrane of the egg, which causes it to swell. Dragonflies have a pointed egg-burster on the head, but damselflies simply push the head against a pre-existing circular line of weakness in the egg-wall. Larvae from the same clutch of eggs may hatch over a period of several days, which probably helps to reduce cannibalism.

The first larval stage, also known as the prolarva, is enclosed in a membrane, which protects it during hatching. As soon as it is free of the egg and has entered the water, it immediately moults. In species that lay their eggs in water, the prolarval stage may last for less than one minute. However, some odonates lay their eggs in damp mud at the water's edge, or in plants or rotten wood adjacent to the water. In these species , the prolarva makes its way to the water by crawling, or by flexing its body to execute a series of jumps, and the prolarval stage can last for up to nine hours.

Larval body form and habitats

The body-shape and colour of an odonate larva reflect the habitat in which it lives. Larvae inhabiting the bottom of ponds and lakes or slow-flowing rivers live among silt and plant debris, often in relatively low light levels, and have small eyes, long legs and antennae, and broad flattened bodies, covered in long fine hairs, called setae. The long legs and flattened body help to prevent the larva from sinking into soft sediments by spreading

RIGHT **The larva of the downy emerald *Cordulia aenea* is a typical sprawler adapted for life at the bottom of the pond. The long legs and broad body help to stop it sinking into mud. It finds its prey by touch using its long antennae.**

its weight: such sediments are often low in oxygen, even a few centimetres below the surface, so the larva could not survive if it were to sink out of sight. Particles of sediment and detritus are trapped in the setae, helping to conceal the larva from predators. Because they often live in rather gloomy habitats, the larvae that live on the river, lake or pond bed have no use for large, well-developed eyes. Instead, they detect their slow-moving prey by touch, using their long legs and antennae, and their clothing of body setae.

Other larvae burrow into the bottom sediment, scooping out a shallow pit with their legs and resting with the top of the head and tip of the abdomen protruding from the sediment. In these larvae the eyes are conical, and project above the general line of the head. Some species can fully conceal themselves within a few seconds of touching the substrate. Others burrow more deeply into the sediment. They have a more or less cylindrical body rather than a flattened one, and a long breathing-siphon, which may be up to half as long as the abdomen and allows them to burrow several centimetres into the sediment. The larva holds its breathing-siphon above the surface of the sediment, to keep itself in contact with oxygenated water and to prevent fine sediment entering the siphon.

By contrast, larvae that live among submerged plants in the water column have large eyes, short legs and short antennae, and streamlined, torpedo-shaped bodies with no covering of fine setae. Light levels are high in this environment, and so these larvae detect prey by sight, using their keen vision. They have no need of long antennae to detect their prey, which is often fast moving. These larvae can be active hunters, or may ambush their prey while lying concealed, clasping the stems or leaves of plants.

ABOVE LEFT **This gomphid larva lives buried at the bottom of muddy rivers in Thailand. Notice the breathing siphon at the tip of its abdomen.**

ABOVE **Aeshnid larvae, which live among plants near the surface, have streamlined bodies so that they can move rapidly through the water.**

Range of habitats

Odonate larvae occupy a wide range of habitats. Most are confined to standing or flowing fresh waters, but a few species are found only in saline or terrestrial habitats. Physiological adaptations have allowed some species to survive in a variety of more extreme environments.

Temperature is the overriding environmental variable governing the distribution and abundance of Odonata. The majority of species occur within the tropical latitudes, and the number of species declines with increasing latitude and altitude. Only one species, one of the emerald dragonflies (family Corduliidae, see also page 80), *Somatochlora sahlbergi*, has its centre of distribution in the Arctic, but 48 species have been recorded as breeding within the Arctic Circle.

Protective colours

The colours of odonate larvae as well as their body shape is usually closely related to their habitat. Most are the same colour as their background, helping them to hide from approaching predators – and prey. Species that live among dead leaves, debris and silt at the bottom are brown or black, whereas those living in plants nearer to the water surface are usually green, and those living in sand are yellow or pale brown. To a certain extent, larvae are able to change colour during moulting. For example, the larvae of one of the American emperor dragonflies, the common green darner *Anax junius*, are mostly green during the summer but brown for the rest of the year. Young aeshnid larvae of many species are marked with a disruptive pattern of dark and pale stripes. These markings appear to protect them from being attacked by other dragonfly larvae since larvae of less active species are not banded and the stripes are lost once the larvae are too large to be eaten by other dragonflies.

Several species occur at high altitudes in tropical regions: the record is held by the globe-skimming dragonfly *Pantala flavescens*, which was recorded migrating at an altitude of 6000 m (19,700 ft) in Nepal, while one of the hawker dragonflies, *Aeshna peralta*, is known to breed in the Peruvian Andes at up to 5000 m (16,400 ft). Some species that have predominately northern distributions also occur in alpine regions in the southern parts of their geographical ranges. Conversely, some tropical species are able to extend their distribution northwards by living in thermal springs. Some of these can even complete their development in the hot springs, surviving temperatures of up to 40°C (104°F).

BELOW **The seaside dragonlet *Erythrodiplax berenice* from the eastern USA is the only dragonfly in the world that is associated with marine habitats.**

Many tropical odonates, including those that breed in seasonally dry pools, will also tolerate saline waters up to the equivalent of 50% sea water. However, the seaside dragonlet dragonfly *Erythrodiplax berenice* is the only odonate associated with the marine environment, breeding as it does in coastal marshes in the eastern USA.

The larvae of most species of Odonata require well-oxygenated, unpolluted water, but there are some that will tolerate conditions of low oxygen. Many species are able to survive in acid water, and will persist in lakes affected by acid rain long after fish and many invertebrates have been eliminated by the increasingly acid conditions.

How larvae move

Odonate larvae spend most of their time motionless. However, they do sometimes move – usually in response to the presence of prey, predators or other odonate larvae – either by walking, swimming or 'sidling'.

Some species, especially those that live among aquatic plants, are quite active, stalking their prey by walking. Damselfly larvae and dragonfly larvae have different methods of swimming. Damselfly larvae flex their abdomens from side-to-side, using their caudal appendages as paddles. When swimming slowly, for example to change perch, they spread their legs wide to act as stabilizers. But they can also swim faster, when they need to avoid a predator, by

ABOVE **This hairy dragonfly larva *Brachytron pratense* keeps the plant stem between it and its enemies. The 'fluff' surrounding the larva are protozoans which are attached to its body by microscopic threads.**

rapidly wiggling the abdomen, and in this case they hold their legs against the side of the abdomen. Dragonfly larvae, by contrast, swim using jet propulsion, by squirting water out of the end of the abdomen. They can achieve very rapid bursts of speed or can hover in midwater.

Odonate larvae that have narrow abdomens and habitually clasp the stems of aquatic plants may escape the attention of predators by 'sidling' – rapidly turning around to the opposite side of the stem. Their bodies are thus concealed from the predator, but their eyes still protrude from the stem so they are able to keep a lookout in case the predator continues its approach.

Larval growth

Insects are encased in a tough, waterproof layer called the cuticle, composed mainly of chitin – a protein similar to the keratin that forms the basis of fingernails and hair in humans. The chitinous cuticle provides a rigid outer-skeleton (or exoskeleton) for the insect, but is inflexible and cannot stretch. This means that insects must have segmented bodies that will flex at the joints between segments, and also means that they must shed the outer skeleton in order to grow. Odonata pass through 8–18 larval stages depending on species, before development is complete.

When a new larval stage is ready to emerge, it becomes separated from the old larval cuticle surrounding it. The larva then pumps fluid by contraction of its abdominal muscles, into its thoracic region, causing it to swell. The outer cuticle splits down a line of weakness running along the middle of the upper surface of the thorax and up to the back of the head before it splits into a Y-shape next to the eyes. The new larval stage can then push itself through the hole in the old larval cuticle. The thorax is the first to emerge, followed by the head and legs, and finally the abdomen is pulled clear.

The new cuticle is soft and very pale because pigmentation has yet to develop, but it is also flexible allowing the new larval stage to increase in size. At this time the larva is very vulnerable to mechanical damage and to predation since it lacks cryptic coloration. However, within an hour the cuticle darkens and hardens.

Development

Moulting is under hormonal and temperature control, with thresholds above or below which it is prevented. As well as changes in overall size and weight, the wing-sheaths, rudimentary genitalia and eye-width increase disproportionately in size. It is possible to deduce the developmental stage of a larva from the number of abdominal segments covered by the wing sheaths. Wing-sheaths are absent in the very early larval stages; they usually extend over the fourth abdominal segment by the final larval stage, or may even cover most of the abdomen in exceptionally squat larvae. The sex of a larva usually can be determined by the presence of a genital protuberance below the second abdominal segment in males or a rudimentary ovipositor below the eighth segment in females. These structures become more obvious as the larva develops.

Most Odonata spend by far the greatest part of their lives as larvae. The larval stage can last from a few months to several years. Larval growth rate is dependent on temperature, the availability of food or seasonal constraints (cold winters in temperate regions or drought in tropical regions).

Duration

The larvae of species that inhabit warm, shallow temporary ponds in dry regions of the tropics have particularly rapid development, which must be completed before the pool or river dries up if the individual is to survive. For example, larvae of the globe-skimming dragonfly *Pantala flavescens*, which frequently breeds in small seasonal pools in the tropics, or larvae of the scarce emerald damselfly, *Lestes dryas*, which breeds in plant-choked ditches that dry out in late summer in Europe and North America, complete larval development in under 50 days. On the other hand, large species living in

RIGHT **The larvae of the blue-tailed damselfly *Ischnura elegans* have a faster rate of development in the warm waters of southern France than in the cool conditions of northern Britain.**

subarctic conditions of cold and low food availability grow slowly. For example, larvae of the golden-ringed dragonfly *Cordulegaster boltonii*, which inhabit cold, nutrient-poor streams in northern Europe, may take five or six years to complete development.

Even within the same species, the time spent in the larval stage can vary, depending on the latitude or altitude at which the individual lives. So, for example, larvae of the blue-tailed damselfly *Ischnura elegans* usually take two years to develop in northern Britain, and only one year in southern Britain. In southern France, the species may have two or even three generations in a single year.

Seasonality

In some species of Odonata, larval development is unregulated, and so the adults could emerge at any time of the year and may be on the wing all year round – or at least for most of the season. Such species are said to have a long flight season. In temperate regions, emergence is usually restricted to the summer months because cool water-temperatures at other times of the year prevent them entering the final stage of development. However, winter emergence has been recorded when a river receives heated water that has been used to cool turbines.

The advantage of a long flight season is that poor weather conditions, which might kill adults or at least restrict their mating success, is unlikely to affect the whole population. The disadvantage is that a relatively small percentage of the population is on the wing at any one time and this may reduce the chances of mating encounters.

To maximize the number of adults on the wing at a particularly advantageous time of year, some species of Odonata have regulated larval development. In temperate species, development is temporarily arrested in a particular stage to enable slow developers to catch up with individuals that have grown more quickly. This arrested development, or diapause, is usually triggered by an environmental cue.

In species living in temperate regions, diapause is triggered towards the end of summer in the year prior to emergence by changes in day-length. Larvae go into diapause as they enter the final or penultimate stage after a critical date determined by day-length. This means that most of the larvae of that generation will be ready to emerge together the following spring. A few precocious larvae that enter the final larval stage before the threshold date will be ready to emerge late in the summer of the first year, thus prolonging the flight season and providing these adults with the potential of mixing their genes with adults of a different year-class.

Feeding

The modified lower lip (labium) of Odonata larvae is a unique feeding organ. This structure has a hinge at its base and another about half way along its length. This arrangement allows the insect to fold the labium away neatly beneath its head when it is not being used to catch prey. At the tip of the labium is a pair of modified sensory organs, called palps, which open sideways and are equipped with a sharp spine at the tip.

ABOVE **The flat labium typical of hawker dragonfly larvae (Family Aeshnidae) can be seen underneath the head of this southern hawker *Aeshna cyanea* larva.**

ABOVE RIGHT **In Libellulidae larvae, like this broad-bodied chaser *Libellula depressa*, the labium resembles a mask covering the lower part of the front of the head.**

In the dragonfly families Cordulegastridae, Libellulidae and Corduliidae the palps are broad, serrated and spoon-shaped, and cover the front of the face of the larvae like a mask (hence the alternative name 'mask' sometimes used for this structure). But in damselflies and other families of dragonflies the palps are narrow and flat, and do not cover the face.

Catching prey

An odonate larva strikes at prey within range of the labium and attempts to impale the prey with the sharp spines at the tip of the palps. Using hydraulic pressure caused by contraction of the abdominal muscles, the odonate larva is able to extend the labium at high speed (less than 25 milliseconds). This movement is so fast that the odonate larva often has time to strike several times before the potential prey has had a chance to take evasive action. The force of the strike is sometimes enough to cause the larva's body to recoil. In the larvae of many species, long spines on the last few abdominal segments may act as a brace against recoil.

Once the larva has successfully caught its prey, it refolds its labium beneath its head and brings the prey up to its 'jaws', or mandibles. The mandibles are very sharp and tough, and are armed with strong teeth. Like the palps at the tip of the labium, they also hinge sideways and shear across each other when closed. Consequently, they can cut easily into the soft flesh of vertebrate prey and even the resilient cuticles of insects.

Dragonfly and damselfly larvae eat a wide variety of prey, including large protozoans, snails, flatworms, and arthropods (including other odonate larvae and ovipositing adults), as well as larval amphibians and fish. They will readily switch between prey, and feed on whatever is most abundant or easiest to catch. Odonate larvae usually attack and consume live, moving prey. Their eyes are specialized to detect movement, but they can also discern shapes, since some species will also scavenge dead and living snails or motionless damselfly larvae. They do not deliberately eat vegetable matter.

Odonate larvae are essentially ambush predators. Most of the time they wait for prey to come to them, sprawling among plants or clasping their stems, or hiding partially buried amongst silt and plant debris. They can hunt and capture prey in complete darkness or in very turbid water, using their long legs and antennae to detect pressure-waves in the water. The larvae may actively hunt prey at night, especially when they are hungry or when there are lots of predatory fish around during the daytime.

Detecting prey

When it detects prey, the larva becomes alert and raises itself from the substrate, pointing its antennae towards the intended victim. As the larva moves its head in the direction of the prey, the antennae converge towards it. The larva then walks towards the prey and bends the antennae to examine it and assess its size. If it deems it suitable, it raises its antennae and holds them wide apart to capture the prey in the labium. After it has eaten the prey, the larva cleans its mouthparts

ABOVE **Dragonfly larvae are ferocious predators. Fully grown hawker dragonfly larvae (Family Aeshnidae) can easily take on small fish like this stickleback (Gasterosteus).**

and returns to its resting position. Larvae using visual means to detect prey have short antennae but large eyes that wrap around the whole head and provide the insect with 360° vision. They can detect prey at least 15 cm (6 in) away – depending on the size of prey and the amount of light available – and turn the head to view the prey with the front of both eyes, since this region has the greatest visual acuity. The view from both eyes intersects at a distance of 5 mm (1/5 in) in front of the head, and this is the optimum striking distance for the labium.

Prey moving slowly across a substrate is stalked and attacked at a distance of about 0.5–1.5 head-lengths away. Snails are attacked at closer range, about 0.2–0.8 head-lengths away. The larva grasps the snail by its foot.

Dragonfly larvae approach swimming prey using jet propulsion, which is accomplished by squirting water from the tip of the abdomen. Relatively large, struggling prey, such as fish, must be subdued. Some dragonflies achieve this by pulling prey into their burrows or beneath the substrate or by stabbing them with the spines at the tip of the abdomen.

Respiration

Aquatic insects must solve the problem of breathing while underwater. Many aquatic insects, especially beetles and some bugs, return to the surface periodically to replenish a bubble or film of air that they carry around with them while swimming underwater. Other insects, including many bugs and larval flies, remain in almost constant contact with the air by means of a long breathing siphon, which sticks out from the tip of the abdomen and breaks through the surface of the water. However, both these strategies leave the insect vulnerable to predators.

Odonata have solved this problem by using gills, which are able to extract oxygen dissolved in the water. Gills are thin-walled outgrowths of the body, which contain a dense, branching network of hollow tubes called tracheae. Oxygen passes out of solution in the water and across the gill membrane into the tracheal system. From there it passes by diffusion, assisted by the pumping action of the abdominal muscles, throughout the body of the larva. Because the water surrounding the gills soon becomes exhausted of oxygen, a constant supply of oxygenated water must pass across the gill surface. In fast-flowing water this is not a problem, and the larva can remain still, but in more sluggish or standing water the larva must create a current itself.

BELOW **The internal gills of a dragonfly larva (left). The thin cross-section and multiple plates ensure that the maximum amount of oxygen is transferred from the water to the gills, from where it diffuses around the body in the hollow tubes of the tracheal system. The branches of the tracheal system terminate in a series of thin gill plates shown on the right.**

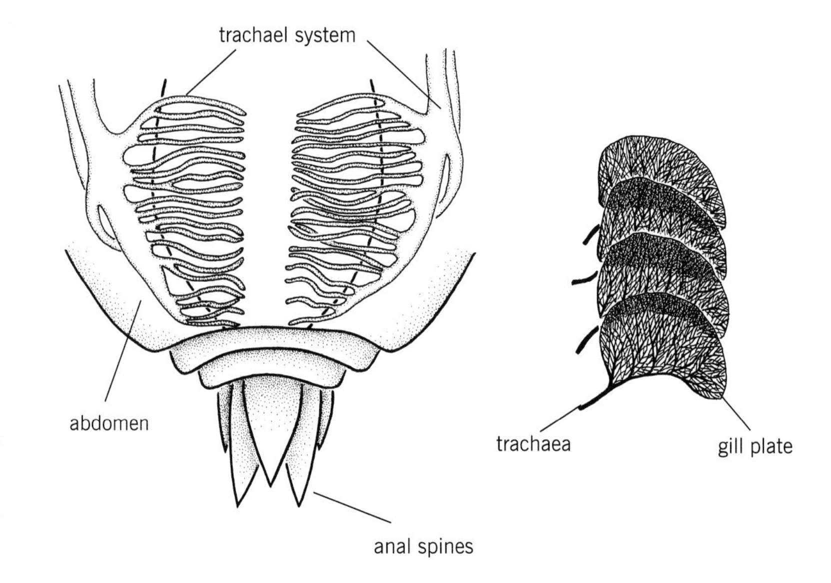

Dragonfly larvae have 30–40 gills inside the rectal chamber of the abdomen, which are elaborately folded and richly supplied with tracheae to form a branchial basket. The larva draws water through the rectum and over the gills by contraction of its abdominal muscles. In conditions of reduced oxygen concentrations, the insect can pump water through the gill chamber more rapidly.

External gills

Damselfly larvae also have rectal gills, but these are not as large or elaborate as those of dragonfly larvae, and so they have additional respiratory organs. The main function of the leaf-like caudal appendages at the tip of the abdomen in damselflies is to serve as gills. In most species, the caudal appendages are broad, with a very thin cross-section. They are well supplied with tracheae. In conditions of low oxygen concentrations the damselfly larva waves the tip of its abdomen from side-to-side to create a current of water. It may also achieve this effect by flexing its legs, almost as if it were doing press-ups.

In high concentrations of dissolved oxygen, damselfly larvae can absorb oxygen across the surfaces of the abdomen, the wing sheaths and the thorax. Consequently, in torrential streams where dissolved oxygen concentrations are high, damselfly larvae are not so reliant on the caudal lamellae to function as gills. Instead, in many species that live in this kind of habitat, the caudal lamellae are relatively small, with a much thicker cross-section and a thicker outer membrane, and may be concave on the lower surface, to act as suction pads that help the larva adhere to the substrate.

The caudal appendages of damselfly larvae may also function as paddles to increase swimming speed, as mechanoreceptors (sense organs that respond to a mechanical stimulus such as touch), weapons or signalling devices. The larvae can also use them to distract predators, shedding them if they are seized. The appendages grow back in successive moults, but in the meantime the damselfly larva must obtain oxygen across other surfaces of its body.

Some groups of damselfly larva have additional gills that function solely as respiratory surfaces. Members of the family Amphipterygidae (see page 68), which inhabit upland streams and rivers in the tropics, have tufts of multi-branched gills situated at the tip of the abdomen. Species in the families Euphaeidae (gossamer-wings; see page 70) and Polythoridae (bannerwings; see page 70) have long, narrow, segmented gills arranged along the sides of the abdomen, but held tucked beneath it to protect them.

When oxygen concentrations are very low, as might occur in warm or nutrient-rich (eutrophic) water, both damselfly and dragonfly larvae can breathe oxygen directly from the air. They move into very shallow water, or reverse up the stem of an emergent plant, until the tip of the abdomen protrudes into the air.

Contests between larvae

Odonate larvae prefer to keep their distance from one another. This makes sense, because they are fierce predators and need to reduce the risk of cannibalism and maximize their success at catching prey.

When a damselfly larva approaches another of similar size, it will turn to face the intruder. Both larvae remain still, staring each other down until one, usually the newcomer or sometimes the smaller of the two, retreats. Sometimes the larva will raise or swing its caudal appendages towards the other one inviting it to leave.

When two dragonfly larvae meet, each might try to stab the other with the spines at the end of its abdomen. If these tactics do not cause an intruding larva to retire, a fight may ensue, in which each larvae strikes at its opponent with its labium. This may result in the loss of a leg or a caudal appendage, but such injuries are not serious in the long term since these appendages can be regenerated in subsequent moults.

The dark markings on the caudal appendages of many species of damselfly larvae serve as a marker to attract the attention of an antagonist and to divert a strike by its labium to this less vital part of the body. These contests are not territorial, but serve to space out the larvae. In small, confined habitats, such as water-filled rot-holes in trees, clashes between larvae of different species often result in the death of one of the rivals.

Metamorphosis

Metamorphosis from the larval stage to the adult damselfly or dragonfly is under hormonal control. During larval development, metamorphosis is inhibited by juvenile hormone. When larval development is complete, increases in day-length and temperature provide cues that trigger the

LEFT **The wings and enlarged eyes of the adult insect can be seen clearly beneath the cuticle of this metamorphosed larva of the North American comet darner *Anax longipes*.**

RIGHT **Larvae of the beautiful demoiselle *Calpopteryx virgo* live in fast-flowing, gravel-bottomed streams throughout most of Europe, except the extreme north and south-west, through Asia to Japan.**

release of hormones promoting metamorphosis and suppressing juvenile hormone. Metamorphosing larvae can be recognized by their splayed wing-sheaths and an increase in the area of dark pigmentation around the eyes. During the final stages of metamorphosis, about a week before adult emergence, the labium contracts, preventing the larvae from feeding.

The rate of metamorphosis is temperature-dependent, and may last from a few days to two months. Metamorphosed individuals gather in the warm shallows of the water, often with their heads and swollen thoraxes protruding above the surface of the water. Their larval gills are no longer functional, and now they must breathe through small openings, called spiracles, in the thorax.

Adults

For a human observer, the process of an adult damselfly or dragonfly emerging from the larval skin is one of the most spectacular natural events that one can witness. Once it has completed its metamorphosis, the odonate is ready to undergo its final moult and emerge. When the larva eventually leaves the water, it is essentially no longer a larva, for the cuticle of the final larval stage conceals the fully formed adult insect.

Emergence

Although moulting larvae are vulnerable to predation, emerging adults are even more exposed. It may be several hours before the insects are ready for flight, and during times of mass emergence birds soon learn that there are easy pickings to be had, and gather at the water's edge to feed on the helpless emerging adult Odonata. For this reason, most large dragonflies emerge during the hours of darkness. Smaller dragonflies and damselflies emerge at dawn among dense stands of reeds and rushes, where they are concealed from predators.

Most species emerge on more or less vertical supports. The metamorphosed larva climbs the stiff stems of emergent plants at the water's edge. Some species emerge horizontally, especially those living in flowing water, the metamorphosed larva climbing out onto boulders at the bank or even in mid-stream.

While most individuals choose to emerge within a few centimetres of the bank, some travel much further and may climb high into bankside trees or walk up to 30 m (100 ft) before settling on an emergence support – after ignoring countless others that appear perfectly suitable to the human eye. The purpose of a few individuals making long

SEQUENCE BELOW
An adult common darter *Sympetrum striolatum* emerging from the larval skin. Emergence may last over two hours and usually occurs on warm, sunny mornings in mid-summer within a few hours of dawn. The head and thorax begin to burst out of the larval skin, when they are free of the larval skin the legs begin to be withdrawn. The hanging-back phase allows the legs to harden. Fluids are

pumped into the wing veins and the wings begin to expand. The green hue of the wings and abdomen during expansion is imparted by the colour of the body fluids. When the wings are fully expanded and the fluids are drawn out of the wing veins and pumped into the abdomen which begins to lengthen. Finally the abdomen is fully extended and is beginning to become pigmented.

journeys before emergence may be to avoid predators that otherwise focus their attention on the bulk of the population emerging near the water.

Having settled on the emergence support, the metamorphosed larva tests its grip by flexing and stretching its legs and flicking its abdomen. This activity also helps the larva to detect any obstructions that may prevent the wings or abdomen of the adult from fully expanding. The metamorphosed larva may settle for many minutes before emergence begins, or may even return to the water to emerge on another day, especially if it is raining.

Once emergence has begun, there is no turning back. The insect pumps fluids into its thorax, causing it to swell and the larval cuticle to split along the lines of weakness on the thorax and head. The adult's thorax is then thrust through the hole and the head, legs and first few abdominal segments follow. The adult may become trapped inside the larval skin if the hole is not wide enough. The abdomen is not pulled clear at this stage, and anchors the emerging adult inside the larval skin. The insect then hangs in a quiescent state while its legs harden.

In damselflies and gomphid dragonflies, the body of the emerging adult faces forwards, but in dragonflies, the abdomen is flexed in a U-shape so that the head hangs backwards over the abdomen. This resting stage may last over an hour, but finally the emerging adult odonatan grasps the emergence support with its legs and pulls its abdomen clear of the larval skin. The shed larval skin is known as an exuvia (plural exuviae).

The adult then pumps body fluids into the hollow wing-veins to expand its wings. The wings are very soft and easily damaged at this stage. The wing-veins may be pierced by sharp leaves or stems of sedges or rushes that are too close to the expanding wings, or by raindrops, or by the claws of other larvae climbing over the emerging adult. If the veins are punctured, the sticky green body fluids may ooze out and harden on the surface of

the wing. Any internal obstruction in the wing-vein will prevent the wing beyond the obstruction from expanding. In both cases, this may prevent the adult from flying.

After the insect has completely expanded its wings, it withdraws fluids from them and diverts them to the abdomen, which then begins to expand. Emergence may take two to three hours in dragonflies and one hour in damselflies, but it may be several more hours before the wing muscles are warm enough for the adult to take flight.

Maturation

Freshly emerged Odonata are called tenerals. They can be recognized by their shiny wing membranes, pale pterostigmata and unpigmented bodies. Several more hours must pass for the cuticle to harden, and several days for the adult pigmentation to develop.

In most odonates, the adult does not become sexually mature for a further one to two weeks. However, in some species that must cope with extreme environmental conditions, maturation may be delayed for several months. In species of the European damselfly genus *Sympecma*, the entire winter is spent hibernating as an immature adult. Several temperate species, especially in the genus *Sympetrum*, sit out the hottest months aestivating (undergoing a period of dormancy during the summer similar to hibernation) as immature adults, and in the tropics damselflies in the genus *Lestes* survive the dry season for months as immature adults in a state of siccatation (a period of dormancy during the hot, dry season).

LEFT **Freshly emerged immature adults are called tenerals. They are characterized by their shiny wings and the light pigmentation of their bodies.**

The maiden flight of the freshly emerged adult odonatan is usually short, and takes it away from the water to the shelter of surrounding trees or bushes. During the time when the adult is sexually immature, it actively avoids water, where it may encounter sexually mature, aggressive males that may inflict physical harm. Instead, immature adults frequent rough grasslands, woodland and the forest interior, where they can feed in peace and may form very large aggregations, or they may undertake long migratory flights.

Vision

In the insect world, Odonata have unrivalled sight. No other insects have larger eyes or eyes that contain a greater number of light-sensitive facets, called ommatidia (see box, opposite). The number of ommatidia that make up the compound eye determines the sharpness of vision. The eyes of dragonflies envelop their heads, providing them with 360° stereo vision. They have full-colour vision and can detect ultra-violet and the plane of polarization of light. They are particularly good at detecting movement.

Over 28,000 ommatidia make up each of the big eyes of *Anax junius*. The greatest number of ommatidia, and the largest ommatidia, are concentrated towards the front and top of each eye. The front of the eye is concerned with monitoring forward flight, and the top of the eye with detection of prey, mates, and rivals silhouetted against the sky. These areas of highest acuity do not reflect light and are visible as black spots, or pseudo-pupils, in the eye. The largest eyes and ommatidia are found in those dragonflies that forage at high speed in dark rainforests or at dusk and dawn (crepuscular activity).

In addition to their compound eyes, all Odonata have three additional simple eyes, or ocelli, consisting of a single lens. The ocelli are located on the top of the head between the compound eyes and are arranged in a forward-facing triangle. The ocelli have poor visual acuity but are sensitive to light intensity. Nerves connect the ocelli directly to the flight muscles, and they provide information about the relative position of the horizon so the insect can continually monitor its flight orientation.

Structure of the compound eye

The compound eyes of insects are made up of a group of units comprising lens and sense cells. Each unit is called an ommatidium (plural: ommatidia). The number of ommatidia in the eye ranges from one, in some species of ant, for example, to tens of thousands, in Odonata. Light is received through the lens and guided via the crystalline cone to the rhabdom, which is the site of photoreception. The rhabdom contains visual pigments that are sensitive to colour. Light falling on these pigments causes the cells to be excited and become electrically active. These electrical impulses are passed to the brain via the optic nerve and perceived by the insect as images. Collectively, the ommatidia produce a mosaic of spots of light of different colours and intensities, which form the image of the object.

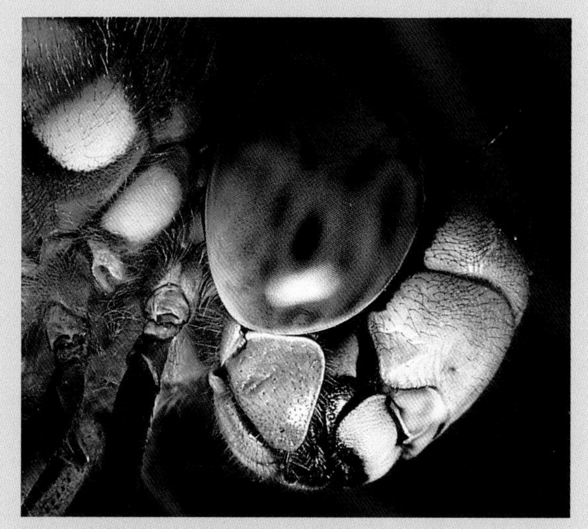

ABOVE **A dragonfly's eyes are made up of thousands of individual lenses (ommatidia), which gives them extremely acute vision. The black spots, or pseudopupils, in the eye of this migrant hawker *Aeshna mixta* correspond to the areas of highest visual acuity where no light is reflected.**

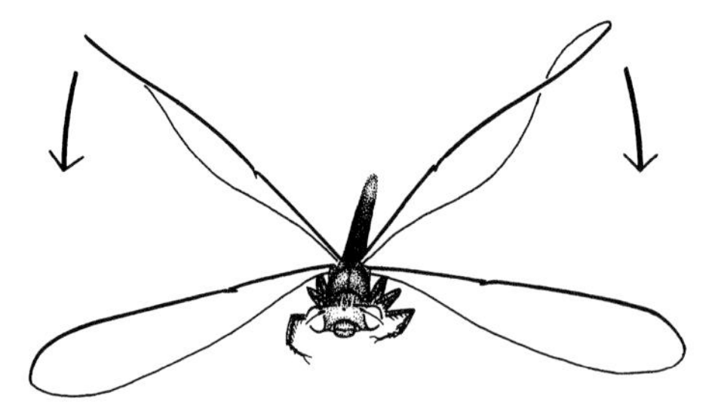

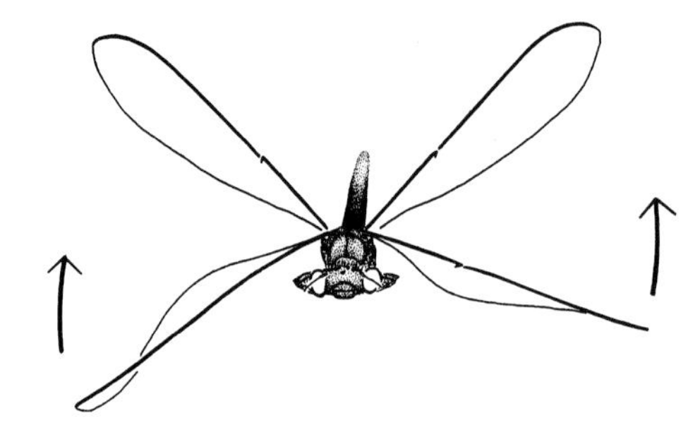

RIGHT **In normal flight Odonata twist each wing in a figure-of-eight. The damselfly on the left is just beginning the downstroke, while the one on the right performs the upstroke.**

Flight

One of the most conspicuous attributes of adult Odonata is their aerial agility and speed of flight. They can fly forwards, sideways and backwards, swoop, soar and hover. This impressive ability is largely due to the power of their large flight muscles and the construction of their wings. The wings are attached to the synthorax by a narrow flexible joint that allows the wings to twist in flight.

Odonata beat their wings independently of each other. In normal forward flight, each wing describes a figure-of-eight motion. The wing is first swept forwards and downwards and then pivoted and drawn backwards and upwards. The wingbeat frequency is relatively low, about 30 beats per second (compared with e.g. small flies and midges which beat their wings at 200–400 beats per second), and the wings do not make a buzzing noise when

RIGHT **A male four-spotted chaser *Libellula quadrimaculata* comes in to land with legs extended and hind wings lowered and acting as air-brakes.**

the insect is in flight. Dragonflies fly at a maximum speed of about 35 km/hr (22 mph) in active, muscle-powered flight, but damselflies can reach only about 10 km/hr (6 mph). Migrating dragonflies which may be on the wing continuously for many hours at a time, conserve energy and reduce body temperature by gliding.

The notch, or node, near the middle of the leading edge of each wing permits twisting and may act as a shock-absorber. The pterostigma (plural: pterostigmata), the thickened, pigmented cell near each wing-tip, may act as a counterweight, promoting wing-twisting and wing-tip rigidity. The shape and density of spines along the leading edge of the wing may act as 'fences', modifying the airflow and reducing turbulence over the wing.

The longitudinal veins of the wing are arranged in a corrugated pattern, with the vein at the leading edge of the wing (the costa) raised relative to the next vein (the subcosta), which in turn is lower than the radius, and so on. The veins are hollow tubes that contain no fluid, and are thicker towards the leading edge and base of each wing. This arrangement provides the wing with extreme longitudinal rigidity, but allows it to twist and flex in the transverse plane. The insect's short antennae are used to detect direction, speed and performance in flight.

Coloration

If the aerobatics of Odonata make them conspicuous, so too do their bright colours. It is not uncommon for their bodies to have

LEFT **A male *Neurothemis stigmatizans* from Malaysia attracts attention with his extensively pigmented wings and white claspers at the tip of his abdomen. The wings of the female are uncoloured.**

RIGHT **Most odonates change colour as they mature. The violet stripes on the thorax of this young female blue-tailed damselfly *Ischnura elegans* will become sky-blue after a few days.**

RIGHT **Female blue-tailed damselflies *Ischnura elegans* have several different colour forms, each of which undergo a series of age-related changes. This female is marked like the male but the green, rather than blue, thoracic stripes indicate that it is young.**

RIGHT **In most populations of the blue-tailed damselfly *Ischnura elegans* a minority of females have a striking orange-red thorax.**

bright blue, red, green and yellow markings, frequently against a black background – often all on the same individual. And colour is not always restricted to the body, since the wings of some species have patches of colour, often at the base, or may be entirely coloured. These coloured areas may be formed from pigments deposited in the wing membrane or body cuticle or, in the case of metallic sheens, may be due to structural attributes causing defraction and reflection of light.

Body and wing coloration is often important in mate recognition, courtship or territorial display, and for this reason the colours of males and females are often different. In many odonate families, males are brightly coloured whereas females are often cryptically coloured. Odonata living in shady tropical forests often have very bright points of yellow or blue on the face and towards the tip of the abdomen that make them shine like lights in the gloom.

The colour of many odonates is age-dependent. Immature male libellulid dragonflies are often the same colour as females. For example, immature males of the common darter *Sympetrum striolatum* are yellow, and become red only when sexually mature. Also, on reaching maturity, many male libellulids develop a grey, white or blue waxy covering, known as pruinescence, over the thorax and abdomen. The eyes of many species change in colour as they mature. For example, in many corduliid dragonflies, the eyes of males change from brown to green. In some damselflies, female may go through a series of age-related colour changes, and five colour forms are known in females of the blue-tailed damselfly *Ischnura elegans*. In old age, the wings of many dragonflies become amber in colour and the females of some species begin to take on the colours of males.

Coloration may also be temperature-dependent in some species. For example, below 16°C (61°F), the abdominal spots of males of the north temperate azure hawker dragonfly *Aeshna caerulea* change from blue to grey, over a period of 40–60 minutes. Similarly, the abdomen of the Afrotropical species *Chlorocypha straeleni* changes from red to grey-black within 25–40 minutes when temperatures fall below 30°C (86°F). Becoming darker in relatively cooler conditions may help these species to keep their bodies warm by absorbing more heat.

Feeding

Odonates are highly efficient aerial predators. They are fast and manoeuvrable in flight. Their large eyes, with numerous ommatidia, are superbly adapted to detect small flying insects against the sky. The thorax is skewed, so that the legs are pushed forwards and can be held beneath the mandibles. And the legs are equipped with long spines on the inner edge that, when held together, form a basket with which prey can be scooped up and gripped.

Odonates will attack, capture and consume any flying insect that is smaller than themselves, especially midges and mosquitoes, and including other species of Odonata. Most are opportunistic feeders, but some species show preferences for particular prey. For example, larger species of *Orthetrum* dragonflies often target butterflies, while the giant helicopter damselflies (family Pseudostigmatidae) of Central America specialize in plucking spiders from the centre of their webs.

Large dragonflies usually consume their prey in flight, but smaller dragonflies and damselflies often land before eating their

ABOVE **This bumblebee's sting was no protection against a hungry golden-ringed dragonfly *Cordulegaster boltonii*.**

RIGHT **Scarlet groundlings *Brachythemis lacustris* gather in the evening at a roost by a sluggish East African river. They orientate towards the fading sun to gather as much heat as possible. Males can be distinguished from females by their bright red abdomens and the orange patch at the base of the hindwing.**

catch. Odonates bite off and discard the wings of insect prey before eating the victim's body. Some species, especially of damselflies, take small insects that have settled on the leaves of plants, a habit that is called gleaning.

Odonates eat up to 20% of their body weight every day. Feeding occurs throughout the day, but is usually focused towards the late afternoon, when large aggregations of mixed species may form to feed on swarms of flies over rough grassland or in woodland glades. Crepuscular species (those flying at dawn and dusk) may gather to feed on insects attracted to lights. Other species may follow cattle, deer or other large herbivorous mammals to feed on the insects they flush from vegetation.

Roosting

Odonates roost during the night or in poor weather, when it is too cold or windy for flight, or during the dry season in the tropics. When roosting, Odonata typically hang down vertically from their perch. Damselflies usually roost close to the ground among grasses, libellulid dragonflies among bushes and aeshnid dragonflies in trees, but all usually select sites that are exposed to the early morning sun.

Sometimes, Odonata roost in large, tightly packed, aggregations (individuals of *Mecistogaster ornata*, one of the giant helicopter dragonflies, actually perch on top of one another!). These include hundreds of mature and immature individuals, of both sexes and sometimes of several different species. These roost aggregations may reassemble each evening, an hour or so before sunset on successive days. At, or just before, sunrise they begin to whirr their wings to warm up their flight muscles and within an hour the roosting adults have usually dispersed.

Arrival at and departure from the roosting site is apparently triggered by light intensity. The same roosting sites may be used for many years, although not by the same individuals, because most adult odonates live for only a few weeks or months, and none for more than a year. At day break, roosting Odonata often raise their abdomens and open their wings to maximise the incidence of the rising sun falling on their bodies in the morning. Crepuscular species often roost during the heat of the day in dense shade.

Temperature regulation

To remain active, Odonata must regulate their body temperature. In hot weather, they are unable to sweat, but they can cool themselves down by dipping into water during flight from time to time. Other strategies for losing heat include perching at the tip of a stem to reduce the amount of heat reflected from below; shielding the thorax by flexing the wings forwards and downwards; pointing the abdomen at the sun (the so-called 'obelisk posture'), which can reduce the body's shadow by up to 50%; and seeking the shade of woodland, caves or overhanging rocks.

Conversely, odonates can increase their body temperature by basking on substrates that radiate heat, such as bare soil and rocks or the trunks of trees; whirring their wings by shivering the wing muscles to generate heat in the thorax and head; and maximizing the proportion of the body that is exposed to the sun by altering posture.

Odonata that live in extremely hot environments are able to tolerate higher temperatures than those living in cool regions. For example, a desert-dwelling skimmer

FAR LEFT **This male hook-tailed dragonfly *Onychogomphus forcipatus* keeps cool in the scorching heat of the Turkish summer by pointing its abdomen at the sun, and shading its thorax with its wings.**

LEFT **The dark colours and thick clothing of hairs on the thorax help to keep this downy emerald *Cordulia aenea* warm in the shady northern European and Russian woodlands it frequents.**

Reasons for migrating

ABOVE **In September 1998 a handful of green darners *Anax junius* were blown off their usual migration route down the eastern coast of North America and turned up in southwest England.**

Long-distance migratory flights of dragonflies can be prompted by several different factors. For example, the vagrant emperor *Hemianax ephippiger* exemplifies a group of species that occur in arid regions of Africa, Arabia and India. They typically occurs in temporary ponds and rivers, and their larvae can tolerate high salinity. The larvae develop very rapidly, in some cases completing development within two months. Large numbers of adults begin to migrate together soon after emergence, and are assisted in their long-distance flights by winds, such as the sirocco that blows into southern Europe from North Africa. This can often carry them to northern latitudes far too cold for them to breed successfully.

Other dragonflies living in temperate latitudes undertake migratory flights to avoid the winter cold. An example of such a species can be found in the green darner *Anax junius*, which occurs throughout North and Central America, China and eastern Siberia. Mature, often mated, adults arrive in the north from further south in early spring, assisted by warm southerly winds. The progeny of these adventurous individuals emerge late in the summer and use thermals (warm air-currents rising from the land) to fly south again before they become sexually mature.

Both the preceding species must migrate each year, to avoid adverse conditions. However, European populations of the four-spotted chaser *Libellula quadrimaculata* undergo mass migratory flights only periodically. In some years hundreds of thousands of individuals, in the pre-reproductive stage, fly together over several days, often following linear landmarks such as roads, canals or railway lines. These mass migrations occur during years of very high population density, and are thought to be triggered by high levels of internal parasites.

Other species may undergo migration at the start of the reproductive phase. This allows them to colonize new sites. For example, during years of high population density, up to 80% of the population of a north temperate species, the black darter *Sympetrum danae*, may migrate following initial reproductive activity.

dragonfly, *Orthetrum ransonetti*, can remain active in shade temperatures as high as 40°C (104°F). The bodies of Odonata that live in cool regions often have a covering of hair on the thorax and basal abdominal segments and are dark-coloured. Some odonates are capable of temperature-induced reversible colour change. For example, in the subarctic azure hawker dragonfly *Aeshna caerulea*, the Tyndall blue pigments become brighter in warm weather, to reflect light and reduce heating of the body.

Dispersal and migration

The powerful flight of dragonflies enables them to cover large distances. Even damselflies can travel hundreds or sometimes thousands of kilometres, using tail-winds to drive them along. Because water bodies are often temporary, since most ponds and lakes eventually fill up with silt and are encroached by trees and other plants, odonates must be capable of finding new breeding sites. However, species that inhabit more permanent water bodies, especially peat bogs or streams in tropical rainforests, tend to be more sedentary.

Pioneer species that prefer to breed in new water bodies that are open and have not been encroached by excessive plant growth, must be good at dispersing. Some such species (especially in the dragonfly families Aeshnidae and Libellulidae and the damselfly family Coenagrionidae) will begin breeding at

LEFT **The broad base of the hindwings of the red saddlebags *Tramea onusta* enable it to glide, and so conserve energy, during its long migratory flights.**

garden ponds shortly after they have been dug. Long-distance dispersal, or one-way migration, is essential in those species that breed in seasonal pools in arid regions of the world, and is also undertaken by some species in temperate regions.

Most Odonata that undertake long-distance migrations belong to the dragonfly families Aeshnidae and Libellulidae, and are characterized by having a large expansion to the base of the hindwing. This enables them to make gliding flights of more than five hours, by increasing the surface-area of the wing for lift, and so conserve energy.

Migratory flights often involve thousands of individuals, which frequently travel at great speed by making use of rain-bearing frontal weather systems. In arid tropical regions, the migrating adults arrive at new breeding localities as the rains begin. These flights may be at heights of over 1000 m (3,300 ft), to take advantage of rising thermal air-currents, and over distances of thousands of kilometres. Dragonflies such as the globe skimmer *Pantala flavescens* and species in the genus *Tramea* are often attracted to the lights of ocean-going ships far out to sea.

In temperate latitudes, several dragonfly species, including the yellow-winged darter *Sympetrum flaveolum*, four-spotted chaser *Libellula quadrimaculata* and green darner *Anax junius*, also undergo long-distance flights, often prompted by high population densities or the arrival of cold fronts. Migrations of temperate species differ from those in the arid tropics in that it is not whole populations that are involved, but they do involve sexually mature adults, as well as immature individuals.

Predators

Odonates are accomplished aerial predators and are the scourge of most other small flying insects. However, they themselves may be taken as prey by a wide variety of predators. The superb aerial agility of adult Odonata is unrivalled, so few predators are able to catch them in flight. However, there are a number of birds that actually specialize in catching Odonata. Foremost among these is the northern hobby *Falco subbuteo*, a smallish, elegant falcon that can spot a dragonfly at a distance of 200 m (650 ft), and may be able to fly at 150 km/hr (93 mph). This dashing predator is able to jink and swoop with a deftness that can match most dragonflies, enabling the bird to grasp them in flight in its talons and consume them while flying by bringing the foot up to the bill.

Many other birds include some dragonflies or damselflies in their diet. These include a number of other birds of prey, including such species as the red-thighed sparrowhawk *Accipiter erythropus* of West Africa and many

LEFT **This rainbow bee-eater was quick enough to catch a migrant emperor (*Hemianax* species) in northern Australia.**

of the falcons, including the Oriental hobby *Falco severus* and the widespread American kestrel *Falco sparverius* and merlin *Falco columbarius*. Hundreds of Swainson's hawks *Buteo swainsoni* may be attracted to feed on large migrating swarms of dragonflies in South America. The purple martin (*Progne subis*) of North America frequently feeds on nothing but Odonata, although most other hirundines (swifts, swallows and martins) take odonates only as occasional prey items. Other birds that catch flying dragonflies include kingfishers (family Alcedinidae), bee-eaters (family Meropidae) and jacamars (family Galbulidae).

Adult odonates, especially damselflies, are also susceptible to predation from other fast-flying insect predators. These include robberflies (family Asilidae), which pounce on the back of passing odonates and can kill even large dragonflies, wasps (families Vespidae and Sphecidae), and also other larger dragonflies. Although helicopter damselflies are able to pluck orb-spiders from the centre of their webs, the spiders are more often the predators, since all but the largest species of Odonata are at risk of being ensnared in their webs or snatched by carefully concealed crab-spiders.

Vulnerable situations

Adult odonates are at their most vulnerable during emergence, and at this time large numbers may be taken by birds such as sparrows (*Passer*) and wagtails (*Motacilla*). The birds quickly learn to gather at the sides of ponds to feed on the helpless insects in the early summer during times of mass emergence. As well as falling prey to birds at such times, Odonata are also at risk from slower moving, ground-based predators, including slugs, ants and hunting spiders.

During oviposition, odonates may be eaten by aquatic predators, especially frogs. In the summer, these amphibians can be the dominant predators of adult odonates, but the insects may also be taken by newts, fish, and surface-feeding hemipteran bugs, such as

BELOW **An azure damselfly *Coenagrion puella* has been caught by a crab spider *Misumena vatia* which was hiding under a leaf.**

BELOW RIGHT **Black ants swarm over a gomphid dragonfly that they have killed while it was emerging. The larval skin lies to the right of the adult's body.**

water-boatmen and water-striders. Parasitic water-mites are often conspicuous on adult odonates, appearing as small red spheres. Mite larvae climb onto teneral odonates during emergence and suck out their body fluids through the cuticle. Once these external parasites have reached maturity, they re-enter the water when the adult odonates return to breed. Adult odonates are often infested internally with parasitic worms and protozoans.

Plants, too, can result in the deaths of odonates. Insectivorous plants, especially sundews *Drosera*, pose a serious threat to odonates that live in bogs. The leaves of these plants are covered in an array of sticky hairs that trap any insects that blunder into them. Once it has caught its victim, the plant releases enzymes that slowly digest its body. Damselflies and weak-flying, freshly emerged teneral odonates are particularly at risk, but ground-perching dragonflies such as skimmers *Orthetrum* and white-faces *Leucorrhinia* can also be immobilized if they are caught by the wing-tips. Odonates can meet an untimely end by accident on other plants, too – especially those with hairy stems or fruits, which can entangle the legs of unwary insects that perch on them. Also, adult odonates are often impaled through the wings by the sharp stems of rushes *Juncus*.

BELOW **Two blue-tailed damselflies *Ischnura elegans* and a small red damselfly *Ceriagrion tenellum* provide much needed sustenance for these sundews *Drosera*, on a nutrient-poor bog in southern England.**

Lifespan

Compared with the length of time spent as larvae, most odonates survive a comparatively short time as adults. In the majority of species, the pre-reproductive period lasts only about two weeks. Most damselflies survive for a further eight days (maximum 64 days) as reproductively active adults, and dragonflies live for about 12 days (maximum 77 days). However, longevity is considerably lengthened in those species that pass through environmentally adverse conditions in the pre-reproductive stage. Species that aestivate may spend up to 18 weeks as immature adults, those that aestivate and hibernate may spend 34 weeks as immature adults, and those tropical species that siccatate during the dry season may spend as long as nine months in the pre-reproductive phase.

Mating

The sex-life of odonates is surprisingly complex. Male damselflies and dragonflies have a wide variety of reproductive strategies that help them to find and secure a mate and to ensure that they are successful at fertilizing the greatest number of eggs that will be laid, in a habitat suitable for subsequent egg and larval development. However, before considering the mating behaviour of odonates, it would be helpful to first consider the adults' unique reproductive organs.

Male reproductive structures

Like most other groups of insects, odonates produce sperm in a pair of testes (his primary sexual organs, or genitalia) located in the ninth abdominal segment. However, male Odonata are unique among insects in possessing a set of secondary sexual organs, located underneath the second and third segments of the abdomen. These include a penis, which the male uses to insert his sperm into the female's body.

This arrangement means that instead of touching the tip of his abdomen to the tip of the female abdomen in order to introduce the sperm to the female genitalia, as in most insects, a male damselfly or dragonfly must adopt a unique copulatory position, known as the 'copulatory wheel' (see also page 55). Before copulation he must first transfer sperm from his primary genitalia to his secondary genitalia: this he achieves by curving his abdomen, so that his primary genitalia are in contact with the secondary genitalia, and he can then pump sperm into a sperm store (the seminal vesicle), which is located under the third abdominal segment. From here, the insect can transfer the sperm to the penis when he is ready to copulate with the female.

The penis

The penis inserts the sperm into the female's body during copulation. In damselflies, the penis, which is located under the second abdominal segment, is rod-shaped and often bears a series of spines. It may also be equipped with a pair of whip-like flagella at the tip.

In dragonflies, the penis has a completely different structure. It is situated below the third abdominal segment and has an inflatable head. When it is introduced into the

LEFT **The hooks and lobes of the male secondary genitalia, underneath the base of the abdomen of this skimmer dragonfly *Orthetrum*, help to locate and grip the female genitalia during copulation.**

female, fluid is pumped into the tip of the penis causing it to swell and unfurl an armoury of twisting sacs covered in short bristles, or long, barbed arms or rigid plates. On either side of the penis, dragonflies have an additional pair of peg-like structures (the hamules) that are used to grip the female around her genital opening.

In both damselflies and dragonflies, the male uses his penis not only used to introduce sperm to his mate but also to remove sperm previously introduced by rival males (see pages 56–7), which is when the spines, flagella, bristles, barbs or plates come in useful.

Claspers

Male odonates have claspers at the tip of the abdomen that they use to hold the female during mating. Damselflies have a pair of upper and lower claspers that grip the upper surface of the female's prothorax. Dragonflies also have paired upper claspers, but the lower clasper is derived from a single plate that may be forked. Instead of holding the prothorax, male dragonflies grasp the female by the head, with the upper claspers fitting behind her head and the lower clasper lying over the front of her head, covering her eyes. The male clasper sometimes scars the eyes of the female, providing evidence that the female has mated. In some gomphid dragonflies, the lower clasper is equipped with formidable spines, which may actually punch holes into the head or eyes of the female. Injuries of this kind are apparently not fatal, since a female may bear several mating scars.

Evolutionary advantages

The secondary sexual organs of males are present in fossil odonates from the Permian period (about 290 million years ago), but absent in fossil Protodonata from the Upper Carboniferous period (325 million years ago). Protodonata have a paired penis near the end of the body, on the eighth abdominal segment, and presumably copulated in the conventional way for insects. The unique secondary genitalia of Odonata may have evolved to allow mating pairs to adopt the precopulatory tandem formation (see also

TOP **The claspers at the tip of the abdomen of hook-tailed dragonflies *Onychogomphus* are particularly formidable.**

ABOVE **The curved blade below the tip of the abdomen of this female southern hawker dragonfly *Aeshna cyanea* is used to cut a slit in plant stems into which an egg is inserted. The style with sensory hairs at the tip can be seen just to the left of the blade's tip.**

page 54), which prevents the male from being usurped by rival males, protects the male from being eaten by the female, and allows the pair to copulate in flight.

Female reproductive structures

The female reproductive organs occupy much of the abdomen, and for this reason the female abdomen is often thicker than that of the male. The genital opening is situated on the underside of the eighth abdominal segment. It is surrounded by the ovipositor, whose structure varies according to whether the species deposits her eggs endophytically or exophytically (see pages 61–3).

Female damselflies, as well as aeshnid dragonflies, have an elaborate ovipositor, comprising a series of three paired valves underneath the eighth and ninth segments. An outer pair encloses a long, curved and serrated blade, which the female uses to cut holes in plant tissue; she uses the inner pair to enlarge the holes. The blade is dark brown because it contains a lot of chitin, which makes it very strong.

At the tip of the valves is a paired finger-like structure, called the style, that the female probably uses to assess the suitability of a substrate for egg-laying. Cordulegastrids possess a fearsome-looking ovipositor that protrudes conspicuously beyond the tip of the abdomen. However, it has a simpler structure than that found in damselflies and aeshnid dragonflies, and it is unserrated.

In libellulid and gomphid dragonflies, which oviposit onto the surface of leaves, stones and so on, the ovipositor is reduced to form a gutter-like plate. The genital opening connects internally with the vagina and a paired oviduct. Eggs are produced continuously in numerous ovarioles, which connect with the oviduct. Above the vagina are the bursa and spermatheca, sacs in which sperm is stored alive for the duration of the female's life.

Mating behaviour

The average life expectancy of sexually mature Odonata is only a matter of a week or two. In this short time it is vital that the sexes maximize their chances of mating success. First they must optimize the likelihood of meeting each other. One way of achieving this is to synchronize emergence, to ensure that most of the population is on the wing at the same time. At some point, every female must visit the water in order to lay her eggs. So, by waiting at a suitable breeding site, a male increases his chances of meeting a female. However, most of the other males in a population may also adopt this strategy, so in order to mate with a female, the male must try to prevent other males getting there first. One way of doing this is to exclude other males of the same species from the breeding site by driving them away. Another strategy might be to find the female before she arrives at the breeding site.

Once the male has located a female, he grips her with his claspers to form the tandem position. In this way can prevent other males from mating with her. During copulation, but before introducing his own sperm, the male can attempt to remove any sperm that the female might be storing from previous

RIGHT **Donaldson's dropwing *Trithemis donaldsoni* surveys its territory from a perch by a sunny pond in southern Africa. From this vantage point it can quickly takeoff to drive away rival males, mate with females or catch prey.**

copulations. Also, during oviposition the male can guard the female and try to prevent other males stealing his mate. We shall now look in more detail at the fascinating strategies and tactics that Odonata adopt to ensure the continuation of their genes.

Finding a mate

Most people interested in natural history will be familiar with the sight of swarms of brilliantly coloured dragonflies and damselflies milling about the edge of a lake at the height of summer. Closer inspection will reveal that the vast majority of these insects are males. They have gathered at the waterside to await the arrival of females. The hundreds of blue damselflies, resembling slivers of sky that have fallen to earth, far outnumber the dragonflies, which themselves form a distinct hierarchy. Dozens of red or powder-blue dragonflies perch among the prominent spikes of rushes and reeds, and make periodic sorties with wings clashing noisily with others of their kind. Overhead, a few very large blue dragonflies make their stately progress, ceaselessly quartering the open water and ferociously attacking others that enter their airspace.

The males of many species of damselfly will tolerate the presence of others of their own species, and show little inclination for aggressive behaviour. For this reason, large numbers can congregate even at small ponds. The same cannot be said of most species of dragonfly, or damselflies in the family Calopterygidae. These odonates can be divided into two principal groups, according to their mode of behaviour: 'perchers' and 'fliers'.

Perchers include most species of Calopterygidae and the dragonfly families Libellulidae and Gomphidae. Male perchers adopt a perch that gives them a good view over a potential egg-laying site, which they defend against rival males and from which they can spot the arrival of females. Members of the dragonfly families Cordulegastridae, Aeshnidae and Corduliidae typify the fliers. In this group, the males tend to patrol a stretch of the bank or an area of the lake, periodically hovering over likely oviposition sites to search for females and drive away intruding males.

As the number of male dragonflies at the water increases, it becomes more difficult for resident males to defend their territories. The territories become smaller and smaller, until eventually the carrying capacity of the water body is reached and no more territories can become established. Males then must adopt different tactics to meet females. They may search for them away from the water and arrive at the breeding site already in tandem. Other males find it almost impossible to break the grip of a male already in tandem. Alternatively, they may arrive at the water late in the day, when successful territory holders have already begun to drift away to feed.

Holding a territory

Odonata exhibit a continuum of territorial behaviour, from no attachment or defence of a site, through brief localization and frequent changing of sites, to a strong and persistent site attachment. Some species, especially in the dragonfly family Libellulidae, show extraordinary fidelity and the same individual may return to the same perch repeatedly day after day throughout the breeding season. High male density, marked aggressiveness between males, and high frequency of female visits all reinforce site attachment.

Before establishing a territorial site, male odonates investigate a water body for potential oviposition sites. The male identifies a suitable site by inspection of the substrate, current speed or quality of aquatic vegetation, and by looking for evidence of the presence of larvae or of ovipositing females. Once a male has selected a site, he quickly gains a competitive advantage over intruding males.

Residence time

The amount of time spent at the territory varies between species. For example, males of an Afrotropical dragonfly, the scarlet darter *Crocothemis erythraea*, spend most of the day within their territories. On the other hand, males of the downy emerald dragonfly

Cordulia aenea, which frequent temperate Old World woodland ponds, usually spend no more than about 20 minutes in their territory before departing of their own accord to feed in woodland clearings. While they occupy the territory, they always successfully defend it against intruders. This kind of behaviour allows several males to share the same territory throughout the day.

Males of the southern hawker dragonfly *Aeshna cyanea* may include several different ponds within their territory and visit each one for 40 minutes or so, returning to the same ponds up to eight times a day over a period of 26 days. The longest recorded period of site fidelity concerns a male of one of the helicopter damselflies, *Megaloprepus caerulatus*, that defended a water-filled rot-hole in a tree in the Panamanian rainforest for 90 days.

Size of territory

The density of males present at a site may in part determine the size of a territory. But physical attributes also have a role. Thus, the territories of most species are fully exposed to

RIGHT **The male scarlet darter *Crocothemis erythraea* spends most of the day perched on a stick within its territory.**

the sun and the resident will leave the territory when it becomes shaded. In tropical rainforests, males often restrict their territories to those patches where the sun can penetrate the tree-canopy to the stream. Landmarks, prominent stands of vegetation or bays along the shoreline can all act as territorial boundaries.

The size of territory can vary enormously, but large species generally hold larger territories than small species. The big aeshnid dragonfly *Hemianax papuensis* may defend a territory of 1800 m^2 (19,375 sq ft), and species of the large golden-ringed dragonfly genus *Cordulegaster* may patrol for hundreds of metres along a stream; whereas the tiny damselfly *Agriocnemis femina oryzae*, which belongs to the family Coenagrionidae (appropriately known as midgets), defends a territory of only 0.2 m^2 (2.15 sq ft).

On patrol

While on his territory, a male spends most of his time searching for females and defending the territory against rival males. Searching behaviour may include watching from a perch, as well as making brief sorties or prolonged patrols. While on patrol, males frequently hover, and change the direction in which they are facing, especially in patches of low light intensity, to investigate likely spots where females may be resting or ovipositing. Patrolling flights may become of longer duration and more frequent after an increase in the number of clashes with rival males.

While patrolling, the odonate maintains a constant height, relatively close to the water surface. Periodic bouts of gliding or dipping onto the water surface may help to keep the insect cool. In species of the blue damselfly genus *Enallagma* (known in North America as bluets), males frequently wait immediately above the surface of the water to intercept females as they surface after underwater oviposition. In general, most males do not feed while they are holding territories.

ABOVE ***Hemianax papuensis*** **is emperor over the largest recorded territory of any odonate.**

Defending a territory

A male will usually only defend his territory aggressively against other males of the same species. However, he will often briefly investigate other intruders into his territory.

Aggressive behaviour can be broken down into four components: approach, threat, fight and chase.

The approach flight is usually direct and from the side or below. Threats may be elicited if the intruder turns to face the defender, whereupon the protagonists face each other, hovering, rising and falling, circling or spiralling around one another; sometimes one will lunge towards its opponent. Complex threat displays are shown by species in the damselflies families Calopterygidae and Chlorocyphidae, which involve stylized flight patterns and displays of the strongly marked wings, legs and abdomens. Perched odonates may threaten intruders by sudden wing clapping or by raising the abdomen to expose brightly coloured patches.

If the encounter is not resolved, fighting may break out and one or both males may try to hold or bite the head or thorax of the opponent, sometimes leading to fatal injuries. Physical fighting represents the simplest form of aggressive behaviour and does not occur in those species that have elaborate, ritualized threat repertoires.

Finally, the territory-holder chases the intruder out of his territory, sometimes pursuing him for tens of metres beyond the territorial limit, although he makes no attempt to catch his rival. In most cases, disputes are settled quickly and the male that believes himself to be resident wins. Protracted clashes take place when the protagonists are unsure of their own or their opponent's residential status. In such cases, it is usually the individual with the largest fat reserves, reflected by its agility in flight, that is victorious. Sometimes, a resident male will attack or even ram pairs in tandem or ovipositing pairs in an attempt to win the female.

Non-territorial strategies

Males that have not been able to establish a territory, perhaps because of high male densities at the breeding sites, tend to adopt different strategies in order to meet and mate with a female. This has been studied closely in the banded demoiselle damselfly *Calopteryx splendens*. So-called 'bank-lurkers' often grasp females while they are passing by. 'Pursuers' are more likely to chase females, whereas other males search for females at roosting sites. 'Sneakers' may take females present in the territory of another male who is otherwise indisposed while chasing away other males, or mating with a different female. 'Stealers' tend to attack tandem pairs and displace the attached male. Finally, 'water lurkers' may search for ovipositing females, and form a tandem with them as they submerge or surface, or will even follow them underwater. Another tactic, adopted by the emerald damselfly *Lestes sponsa* and the marshland darter dragonfly *Sympetrum depressiusculum*, is to form tandem linkages with females at the roosting sites and escort them in this way to the oviposition site.

Mate recognition

Males initially recognize females of their own species using a combination of visual cues, including colour, size, shape and flight style. However, they occasionally make mistakes,

RIGHT **The male beautiful demoiselle *Calopteryx virgo* (above), demonstrates his suitability as a mate to a female (below), or sees off rival males, by ritualized display of his wings.**

particularly in conditions of high male densities, and may grasp females of different species, males, or even dead females floating on the water surface or trapped in spiders' webs.

Unreceptive females are able to dissuade males by a downward curve of the abdomen or by darting into dense vegetation. Females can prevent males from forming a tandem by stretching their fore legs above the prothorax. Perched odonates can put off attentive males by clapping their wings or raising the abdomen.

Courtship

In a few species of Odonata, mostly in the damselfly families Calopterygidae (demoiselles) and Chlorocyphidae (jewels), copulation is preceded by an elaborate courtship display, undertaken mainly by the

RIGHT **This male ruddy darter *Sympetrum sanguineum* has misread the cues and is attempting to mate with a female common darter *Sympetrum striolatum*.**

male for the benefit of the female who can choose to accept or reject him. There is a very marked sexual dimorphism (difference in appearance) in species of these families: males and females have different body coloration, and the wings of males are marked with conspicuous dark pigmented patches covering large areas, whereas those of the females are unmarked. The primary function of the male's courtship display is to demonstrate the merits of the oviposition site within his territory.

In Calopterygidae, the first part of the display is initiated when a female flies into the territory of a male. The male draws the attention of the female to the quality and location of the oviposition site by alighting on the surface of the water and floating downstream for a few metres, to demonstrate the current speed. He spreads and flutters his wings, and curls the tip of his abdomen upwards to reveal a patch of bright colour on the underside. The female demonstrates her acceptance of the male by perching in his territory and clapping her wings. The male then hovers in front of her, displaying his wings by increasing their beat frequency, while facing her and moving slowly around her in an arc. After several seconds, if the female has not indicated her refusal by repeatedly spreading her wings, the male lands on her wings or thorax, and forms the tandem link by grasping her pronotum with his claspers.

Courtship display in Chlorocyphidae is broadly similar to that of Calopterygidae, the differences reflecting differences in the

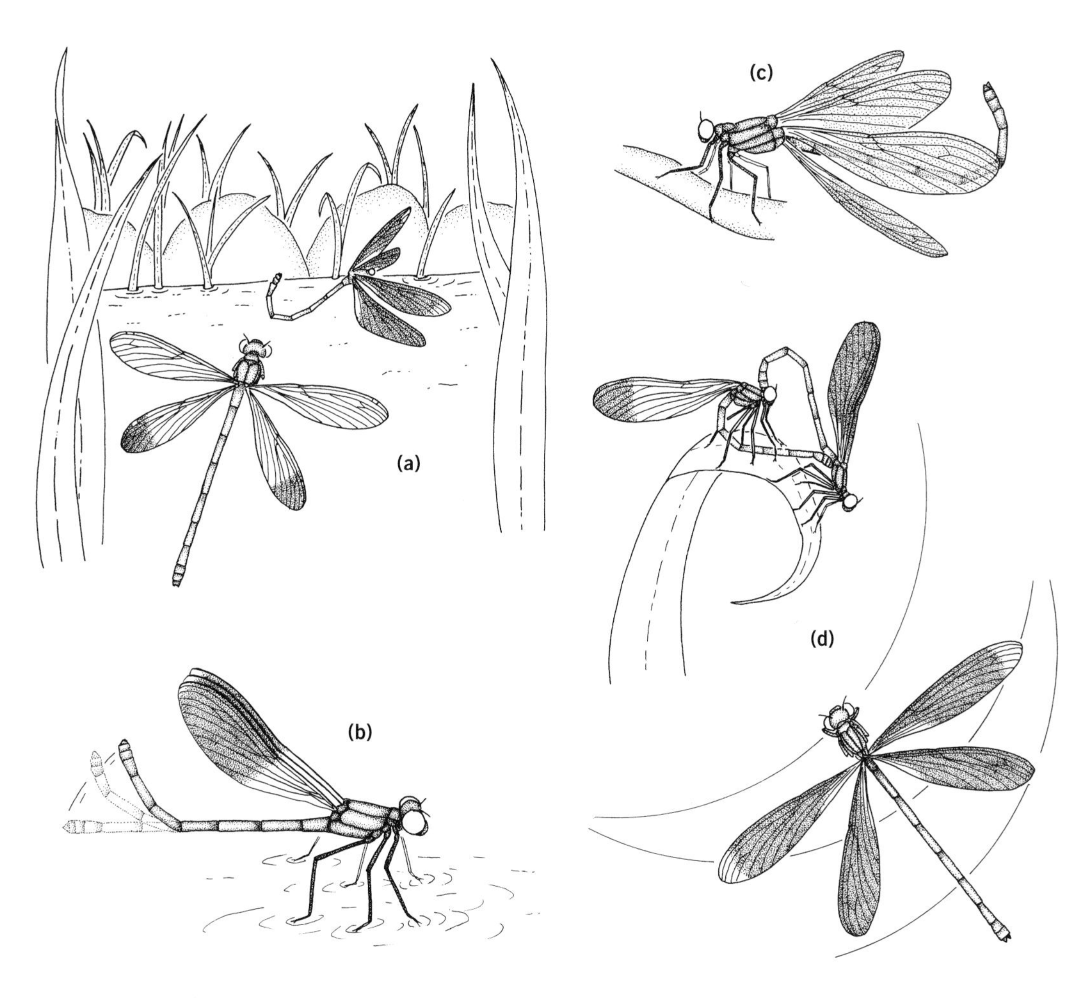

LEFT

(a) Courtship rituals in the Mediterranean demoiselle *Calopteryx haemorrhoidalis*. The dark-winged male (top) leads the female to a suitable oviposition site, guiding her with the bright carmine patch on the underside of the tip of his abdomen.

(b) A male *Calopteryx xanthostoma* demonstrates the suitability of the current speed of the stream in his territory by landing on the water and sailing downstream. He attracts the attention of a female by curling his abdomen to reveal a pale yellow patch at the tip, and waving the abdomen.

(c) The male beautiful demoiselle *Calopteryx virgo* attracts the attention of females from a perch by displaying a pale pink area at the tip of the underside of the abdomen.

(d) The Mediterranean demoiselle *Calopteryx haemorrhoidalis* courts a female by fluttering his wings and moving slowly in an arc in front of her. If the female accepts him she will allow him to land on her and form a tandem.

habitats that they frequent. Chlorocyphids inhabit fast-flowing, upland streams and usually oviposit in dead driftwood. Males and females test the suitability of the wood by a stylized 'treading' movement. During their courtship of females, the males display their iridescent wings by vibrating them and holding them towards the female. In males of some species, the tibiae are brightly coloured and grossly expanded, and the male shakes his legs and displays them to the female as he circles around her. When flying towards the oviposition site, the male allows his brightly coloured abdomen to hang down and swings it from side to side. The female signals her acceptance of the male by perching near the oviposition site with wings closed.

Success in courtship in both Calopterygidae and Chlorocyphidae is related to the ability of the male to choose an oviposition site that is likely to ensure survival of the eggs, and his ability to choose successfully and defend a territory that includes such an oviposition site.

RIGHT **A pair of ruddy darters *Sympetrum sanguineum* in tandem. The male (above) grasps the head of the female with the claspers at the tip of his abdomen.**

The tandem position and copulation

Following visual recognition of a mate, and acceptance of the male by the female after courtship, the male attempts to make a tandem bond with the female. This is usually successful only if there is good physical linkage between the male's claspers and the sculpturing of the female's pronotum, in the case of damselflies, or the rear of the head, in the case of some dragonflies. After habitat selection and visual communication, this form of physical communication is the next tier of reproductive behaviour that helps to prevent interspecific or homosexual pairing.

Securing a tandem linkage is usually rapid, and can occur when the female is in flight or perched. It is preceded by the male grasping the female's wings or thorax in his legs or mandibles. By maintaining tandem linkage, the male can reduce the chances of other males mating with his partner until she has completed oviposition; he can rescue her from predators; and it provides stability during flight, copulation and oviposition. It is during pre-copulatory tandem linkage that the male normally transfers sperm from the primary to the secondary genitalia.

Having formed a tandem, the male then attempts to induce the female to mate with him. Copulation usually follows the formation of a tandem pair within a few minutes. If the female appears to be unreceptive, the male may swing her forward and grasp the tip of her abdomen in his legs and begin to massage it. If even this fails to induce the female to mate, the male will release her.

Because the secondary genitalia of male odonates are situated beneath the base of his abdomen, mating pairs must adopt a unique 'wheel position' during copulation. The wheel can be formed – and the whole copulatory process completed – in flight, although in many dragonflies copulation is completed on the ground or while perched, often away from the water and the disturbance of other males.

To achieve the wheel position, the male bends his abdomen to bring the head of the female forward. She in turn bends her abdomen forwards and upwards until her genitalia, below the eighth abdominal segment, interlock with his secondary genitalia. The pair may also grasp each other's abdomen in their legs to assist in this manoeuvre. To help the female grasp the male abdomen, his hindwings are cut away near the base in all dragonflies except members of the family Libellulidae. The legs of copulating females may rub off the waxy secretion (pruinescence) on the surface of the abdomen

BELOW **A pair of helicopter damselflies *Mecistogaster rotundata* from the forests of Ecuador in a 'copulatory heart'.**

RIGHT **The dark diagonal stripes, half way down the abdomen of this male scarce chaser *Libellula fulva,* were caused by the legs of a female as she grasped him during mating and rubbed off the pale blue waxy covering.**

of male libelullids, leaving telltale patches where the darkly pigmented abdomen shows through from below.

Sperm competition

After a mating pair has formed the wheel position, there is one important task for the male to undertake before he transfers his sperm to the female. This is to remove or displace any sperm that may have been deposited by other males in the female's sperm-storage organs (the bursa and spermatheca) during previous matings. He achieves this by one of the following mechanisms, depending on the species. Males that have hooks or barbs on the penis physically remove the sperm. In other species, the male can reposition rival sperm within the female storage organs, so that the sperm he introduces is the most readily available for fertilization of the eggs. Another tactic is for the copulating male to flush out and dilute rival sperm with his ejaculate. In some species, the male may induce the female to contract her sperm-storage organs to eject sperm. Following sperm removal or displacement, the male then inseminates the female with his own sperm.

Sperm present in the female bursa has priority in fertilizing eggs, and chiefly contains sperm from the most recent mating. Sperm stored in the spermatheca is derived from earlier matings. Females usually store sufficient sperm from a single mating to fertilize several clutches of eggs without the need to mate again for several days.

About 95% of the eggs laid immediately after copulation will have been fertilized during the most recent copulation. But this value declines as the amount of time after the last copulation increases and sperm in the storage organs becomes increasingly mixed. The duration of copulation is directly correlated with the amount of time the male spends removing or displacing sperm. Sperm removal takes considerably longer than sperm displacement.

By spending longer removing sperm deposited during previous matings, a male can maximize the number of eggs laid that were fertilized by him; as long as he can prevent the female re-mating before she has laid the majority of her eggs. However, the trade-off is that this reduces the number of matings he can have with different females. A males that presides over a large territory also stands a greater chance of sperm precedence, because a female with which he has mated is likely to spend more time in the territory and therefore to lay a greater number of eggs before she encounters other males.

The amount of time that territorial dragonflies spend in copulation is inversely related to the amount of time they spend in their territory. Species such as the downy emerald dragonfly *Cordulia aenea*, which remain for only about 20 minutes in the territory, spend at least 1 1/2 hours in the copulatory wheel position. On the other hand, species such as the scarlet darter dragonfly *Crocothemis erythraea*, which shows high fidelity to its territory, may spend just a few seconds in copulation.

Many species of damselfly have protracted copulation that may last several hours. The blue-tailed damselfly *Ischnura elegans* has the longest recorded, and may spend over seven hours in the copulatory wheel position. However, it actually spends a relatively small percentage of this time (about one hour) in sperm removal and insemination. Most of the time, the male is inactive. Delaying oviposition until late in the day and spending protracted periods in the copulatory wheel position is likely to be a form of copulatory guarding and a means to prevent other males from mating with the female.

BELOW **A pair of blue-tailed damselflies *Ischnura elegans* may spend over seven hours in the mating posture.**

Egg-laying

The male odonate goes to a great deal of trouble to fertilize as many as possible of the eggs a female lays. He attempts to select an egg-laying (oviposition) site that is favourable to the survival of the eggs and larvae that he has sired. He guards this site against the incursions of other males. He may court females that enter his territory to demonstrate his suitability as a father, and also the suitability of his territory. He holds receptive females in tandem linkage to prevent other males from taking her before he has mated. He removes the sperm of rival males from the female. But his efforts are not yet complete. He must persuade the female to oviposit in his territory, to lay as many eggs as quickly as possible, and prevent other males from snatching her and mating with her before she has completed oviposition.

Guarded oviposition

Most species of male damselflies and libellulid dragonflies, and a few species in the dragonfly families Aeshnidae, Gomphidae and Corduliidae, guard the female for at least some, if not all, the time during oviposition. This may take the form of contact or non-contact guarding. The intensity of guarding (the proportion of time spent by a male in contact or non-contact guarding) reflects a trade-off between preventing other males from mating with the female with which he has just mated himself, and the availability of other receptive females with which he may mate. This is a function of male and female density.

Within a few minutes of the completion of copulation the pair start oviposition. The male either leads the female to the oviposition site, all the while remaining in tandem, or he may release the female and escort her to the oviposition site, indicating its position by down-curving his abdomen. In some libellulids, the female takes a rest-period between the end of copulation and the beginning of oviposition. She may use it to sort the sperm, though during this time the male may ram her, bite her or land on her, in an apparent attempt to induce her to start oviposition.

LEFT **The males of *Pseudagrion niloticum*, like most damselflies, remain in tandem with the female throughout oviposition.**

RIGHT **This female ruddy darter *Sympetrum sanguineum* has been left by the male to oviposit alone. She hovers and flicks her abdomen to release the eggs which can be seen falling to the water.**

In some species, the male remains in tandem with the female throughout the period of oviposition, which may last for several hours in some damselflies, but is generally shorter in dragonflies. In some libellulid dragonflies, the time spent in tandem with the ovipositing female is dependent on the number of males and receptive females present. If the pair suffer a high degree of harassment from other males during oviposition, then the pair will remain in tandem. If, however, the pair have received little or no interference, then the male may release his grip on the female. At first, he will remain hovering close to the female, ready to drive off any intruding males. At the same time, he will be in a better position to observe the presence of any other receptive females in the neighbourhood. As time passes, he will gradually move further away from the ovipositing female, until she is effectively ovipositing alone.

RIGHT **A perched male mosaic sylph *Chlorolestes tessalatus* stands guard over his mate as she lays an egg into the stem of a waterside plant.**

In those species of damselfly that remain in tandem throughout oviposition, the male often adopts the 'sentinel position'. He holds his body erect and almost vertically above the female, with his legs folded against his thorax. In this way, he is able to prevent other males mating with the female. He can also survey the area for potential predators, either above or below the water, and pull the female clear if any threats arise.

If the female completely submerges during oviposition, the male will release her but wait at the surface for her reappearance, whereupon the pair resume the tandem position. This is vital for the survival of the female, since she is unable to break free of the water surface without the assistance of the male. Sometimes, the female finds herself abandoned on the water surface, but she can use her wings as paddles, and may be able to row herself to the shore or to some emergent support up which she can climb to safety.

In conditions of high male density, the male may descend underwater with the female. In these circumstances, single males may sometimes descend to search for unaccompanied females ovipositing underwater.

BELOW **Male white-legged damselflies *Platycnemis pennipes* remain on-guard in the 'sentinel' position, ready to pull their ovipositing mates clear of the water should danger threaten.**

Non-guarded oviposition

In most dragonflies, with the exception of the family Libellulidae and a few species from other families, and in many species of the damselfly genus *Ischnura*, the males leave the females to oviposit alone. This they do during times of minimum male activity, either early in the morning, late in the afternoon or in overcast conditions. If approached by males, they demonstrate their unwillingness to mate by performing conspicuous refusal displays. Females of the European blue-tailed damselfly *Ischnura elegans* are aggressive towards males or females that trespass on their oviposition sites.

The adaptive advantages of non-guarded oviposition are hard to understand. The male may lose his ability to ensure that it his sperm that are fertilized, and the female is at an increased risk of death when ovipositing alone. But most of those species that adopt this tactic receive sperm in clumps, rather

ABOVE **An African primitive, *Tetrathemis polleni,* coating a twig with a batch of eggs. This species is one of the few libellulids in which the female oviposits alone.**

than free-swimming sperm, and these sperm clumps are not available to fertilize the eggs until several hours after insemination.

Egg-laying strategies

Odonata adopt two basic kinds of oviposition strategy. All damselflies, and also members of the dragonfly family Aeshnidae, practice endophytic oviposition: in other words, they lay their eggs within the stems and leaves of plants. All other dragonflies lay their eggs exophytically, by dropping them in a mass directly into the water, or in a few cases, epiphytically, onto plants.

Endophytic oviposition

Before a female odonate can lay her eggs endophytically, she must land on the plant and grasp it firmly with her legs, in order to obtain sufficient purchase to make an incision in the plant tissue with the long, curved and serrated blade of her ovipositor (see page 45). She lays her eggs one at time, and also pushes them singly, deep into the slit she has cut into the plant tissue. This process is time-consuming – the female can lay only 1–18 eggs per minute – and so it may occupy the female for hours, making her vulnerable to predation. Nevertheless, this form of oviposition is obviously effective, since odonates have practised it for at least 25 million years.

A female rarely lays all her eggs into one plant, and may not lay the entire clutch in one session. Several days may elapse between egg-laying bouts. The eggs are often laid in a characteristic fashion, perhaps in a spiral, in concentric circles or in a line. Plants may form a gall around the egg, and those that receive large numbers of eggs may be killed. The female may choose to oviposit in the stems and leaves of emergent or floating plants, or she may oviposit in submerged plants, sometimes descending up to one metre below the surface and spending up to an hour submerged. While underwater, the female retains a bubble of air trapped between her thorax and forewings. Oxygen passes from the water into the bubble, and a current of oxygenated water is maintained as the female undulates her abdomen during oviposition or flexes her legs. Her forewings are covered by the hind wings, so they remain dry. It is important that they do not become waterlogged if the female is to have any chance of taking off once she has re-surfaced. Having completed oviposition, the female can ascend rapidly to the surface simply by letting go of the plant.

Some species favour particular species of plants in which to lay their eggs. For example, the green hawker dragonfly *Aeshna viridis*, from northern Eurasia, favours the fleshy leaves of the water soldier *Stratiotes aloides*, whereas the white-legged damselfly *Platycnemis pennipes*, from central and eastern Europe and the Near East, often chooses small buds of the yellow water lily *Nuphar lutea*. The Afrotropical virginal spreadwing damselfly *Lestes virgatus* uses the grass *Imperator cylindrica*, which grows in depressions that later fill with water. The European and Near Eastern hairy dragonfly

ABOVE **An unguarded female blue-tailed damselfly *Ischnura elegans* lays eggs into a reed stem just below the water.**

RIGHT **The virginal spreadwing *Lestes virgatus* prefers to lay eggs in dry depressions in which the grass *Imperator cylindrica* is growing. The male is shown in this photograph.**

Brachytron pratense, on the other hand, usually oviposits into the dead leaves and stems of floating reeds, sedges and rushes. The Eurasian red-eyed damselfly *Erythromma najas* oviposits by preference on the underside of large floating leaves of aquatic plants, while the willow emerald damselfly *Lestes viridis* of central and southern Europe and North Africa, lays her eggs just below the bark of waterside trees and shrubs.

Exophytic and epiphytic oviposition

These forms of oviposition are practised by most species in the dragonfly families Gomphidae, Libellulidae and Corduliidae. In these families, the female genital opening is covered by a simple scoop-like plate, in which an egg mass accumulates. Oviposition rates are much higher than in endophytic oviposition. The number of eggs released at once varies considerably between species, ranging from less than five to over 100. Oviposition rates of 100–1500 eggs per minute have been recorded.

During exophytic oviposition, it is usually unnecessary for the female to land. The female flies low over the water surface, trailing her abdomen, or hovers above the water, repeatedly and rapidly tapping the tip of her abdomen on the water surface to wash off the eggs. In some species the female flicks her eggs into the water without coming into contact with it. In several species, this may be facilitated by the female scooping up drops of water, which she retains between short flanges that extend below the abdomen towards its tip.

Odonates that practise epiphytic oviposition place their eggs onto the surface of emergent plants or onto plants close to the surface of the water. Usually, they do this in flight, by swooping down or hovering over the plant. Nevertheless, some species land on the plant during epiphytic oviposition. For example, *Procordulia grayi*, an emerald dragonfly from the South Pacific employs three modes. The female may float on the water surface, perch at the water's edge with the tip of her abdomen submerged, or briefly dive into the water.

BELOW **The flanges visible beneath the eighth abdominal segment of this female roseate skimmer *Orthemis ferruginea* are used to scoop up and hold a few drops of water, which is useful when the eggs are flicked off during oviposition.**

RIGHT **The long ovipositor of this female golden-ringed dragonfly *Cordulegaster maculata* enables it to push eggs deep into the gravel of the shallow streams.**

Although dragonflies in the family Cordulegastridae and a few species in the family Corduliidae have a large ovipositor, it lacks a cutting blade, so these dragonflies are incapable of laying endophytic eggs. Instead they use their ovipositor to insert their eggs deep into soft substrates, such as moss, mud, gravel or between pebbles in shallow streams. Female cordulegastrids hover, unaccompanied by males, above the substrate and pile-drive their eggs into it with rapid and repeated stabbing motions of the whole body. Females of the corduliid genus *Somatochlora* must land on mossy mounds at the edges of ponds to thrust their eggs deep inside.

Some odonates practise exophytic oviposition even though they are equipped with a functional cutting ovipositor. For example, females of the Central American giant damselfly *Mecistogaster martinez* (Family Pseudostigmatidae) hover above water-filled tree rot-holes and catapult their eggs into the water. The Gynacanthini are a tribe of very large tropical dragonflies in the family Aeshnidae with crepuscular habits. Females have a stiff forked structure protruding from the tip of the abdomen. The dragonfly uses this to scrape holes in mud and anchor the abdomen before she drives the ovipositor into the substrate. Sometimes up to one-third of the abdomen may be pushed beneath the surface.

RIGHT **This African forest hawker *Gynacantha villosa* uses the fork beneath the last abdominal segment to scrape a hole in the mud and anchor her abdomen during oviposition. Notice that the tip of one of the claspers was broken when the abdomen was forced into the hard substrate during oviposition.**

Habitat selection

The choice of oviposition sites reflects the situation in which the eggs are most likely to develop successfully and, usually, the habitat in which at least the early larval stages are likely to survive. The larvae of non-migratory species that inhabit temporary pools, or pools and streams that undergo large fluctuations in water level, often lay their eggs some distance above the present water level, burying them in mud or deep within vegetation. This is in anticipation of the water level rising later in the season, and ensures that the eggs do not hatch until there is enough water in the pool or stream for the larvae to have a chance of completing development.

Most species lay their eggs in shallow water or on plants near the water surface, where higher temperatures can promote egg development. Likewise, current speed can be critical: too slow and the eggs may not receive enough oxygen, too fast and the larvae might be washed downstream into unsuitable habitats. Water that has a low pH or supports acid-loving plants may be an indication that the habitat is free of predatory fish.

Recognition cues

Habitat recognition cues are probably hierarchical. The first or highest level is likely to be recognition of a broad habitat type. For example, odonates are able to detect water from relatively high altitudes due to its reflective surface and the insects' ability to visualize polarized light. The presence of woodland, forest or open pasture or moorland, which may relate to availability of prey, water chemistry and water temperature, will also be apparent.

Next, the general shape and size of the water body is also likely to inform the odonate whether it is standing or flowing. For example, flowing water bodies are most likely to be linear; fast-flowing water bodies are likely to be narrower than slow-flowing ones. Dragonflies are sometimes fooled into ovipositing onto the surface of wet roads. A water body with a large surface area is likely to be deeper, cooler and less vegetated than a water body with a small surface area.

Closer visual inspection is the next level, and will enable the insect to distinguish parts of the water body that are unshaded or shaded by bankside vegetation, areas that include dense stands of emergent vegetation, and sections that are well-provided with floating or submerged plants, as well as those that contain no plants. This information may give clues concerning the type of substrate available for larvae at the bottom of the water body, or the amount of cover available to larvae in which to hide themselves from predators.

BELOW **Odonates use a hierarchy of visual and tactile cues to recognize a suitable habitat.**

The final level of recognition, that of choosing the actual oviposition site, may depend on a variety of visual, tactile and possibly chemical information. This might concern current speed, which relates to the concentration of dissolved oxygen in the water, the risk to the larva of being swept downstream, and the particle size of the substrate. The insect may identify critical plant species, whose presence may reveal information about the water's acidity or alkalinity and its overall quality, and the plant architecture of the larval habitat. Assessing the water temperature could help the insect determine the likely development rate of eggs and larvae. Also, the presence of other ovipositing females or the courtship display of males of the same species would indicate the acceptability of the habitat and a reduction of predation risks.

Some species occur in a wide range of habitats, and may be found almost anywhere. Others are restricted to standing or flowing water. Some are restricted to particular habitat types, such as lowland or upland rivers, acid pools or seepages, fens or woodland ponds, temporary pools, dense forest or open meadows. Within these categories, some species are even more discriminating. For example, the Northern emerald dragonfly *Somatochlora arctica* of Eurasia (in the family Corduliidae) will breed only in small bog pools containing mats of vegetation that break the surface of the water, causing sparkling reflections. Other species oviposit only in water-filled rot-holes or on the faces of waterfalls. Becker's pinfly dragonfly *Roppaneura beckeri* (family Protoneuridae), from Brazil, oviposits only in the stems of an umbellifer, *Eryngium floribundum*, and the larvae of this species occur only in the water-filled leaf-axils of this plant.

Fecundity

The number of fertile eggs produced by a female in her lifetime is dependent on a number of factors. These include lifetime egg production, the number of eggs in each clutch, the total number of clutches, the lifespan of the female, the time that elapses between clutches, and the number of mates. These factors are influenced by predator density, mating and oviposition opportunities, and also by the weather – which is very significant in temperate latitudes, particularly the temperature, rainfall and amount of sunshine.

As long as a female is uninterrupted during oviposition, she will normally lay all her available eggs. She will then not have the next batch ready to lay until one to five days later, depending on the species. The clutch size may vary between species, and between individuals of the same species. It is also influenced by the substrate into which the eggs are laid, the mode of oviposition, the size and age of the female, the egg volume, the interval between clutches, the temperature and the presence of parasites.

ABOVE **The record for the greatest number of eggs laid at once is held by the four-spotted chaser *Libellula quadrimaculata*, which can lay over 3300 eggs in a single clutch.**

The four-spotted chaser dragonfly *Libellula quadrimaculata* has been recorded as laying over 3300 eggs in one clutch, but 1500 is more typical for exophytic species, and 500 for endophytic and epiphytic species. However, the weather has an overriding influence on all these variables, and the fecundity of a female can vary by a factor of ten, depending on the number of days of sunshine during her lifetime.

Dragonfly and damselfly diversity

This section briefly surveys the appearance, distribution and behaviour of each of the world's 29 familes of Odonata. Together with the illustrations of a typical species in each family, this will give you a good idea of the impressive biodiversity of this ancient group of insects.

Suborder Zygoptera (damselflies)

Amphipterygidae

Only a handful of species are known in this family, which has a scattered distribution in Central and South America, South-east Asia, and tropical West Africa. Most live in mountain streams in tropical forests. The larvae are unusual among Odonata in having a tuft of gills protruding from the tip of the abdomen.

Calopterygidae (demoiselles)

This family includes many species that live in temperate and tropical regions throughout the world. They are among the largest and most attractive of all damselflies. The bodies and

LEFT **A male beautiful demoiselle *Calopteryx virgo* (Calopterygidae) in Brittany, northern France. This species lives on tree-lined streams with gravel beds.**

broad wings of most species are brightly coloured in iridescent and metallic shades of blue, green or red. The wings of males in many species are also intensely coloured, and are used in courtship and threat displays. These damselflies typically occur among the dappled light and shade found along rivers and streams flowing through forest and woodland. The adult males select territories where pools of sunlight penetrate the tree canopy, and large numbers may gather, perched on the tips of leaves overhanging the water. Calopterygids beat their wings quite slowly, giving them a rather fluttering, bouncing flight.

Chlorocyphidae (jewels)

Chlorocyphids are restricted to the Old World tropics. They have a rather stocky appearance, with long narrow wings and distinctly elongated, snout-shaped faces. The bodies of males are brightly coloured red, blue or yellow, or sometimes in a combination of these colours. The wings are often marked with black patches, leaving small 'windows', which have a pink or blue iridescence in the sunlight.

These damselflies breed in fast-flowing forest streams, and can be seen perching on boulders and overhanging plants or flying fast, close to the water surface. The males of several species have broadly flattened tibiae that are brightly coloured red or white, and are used to semaphore a threat or courtship display while the insect is in flight.

Dicteriadidae (barelegs)

This family contains only two known species. Both occur in streams in the Amazonian rainforest, and both are rarely seen. They are unique among Odonata in having long thin legs that lack long spines, which suggests that they may not use their legs when capturing their prey.

TOP **A male *Chlorocypha curta* by a fast forest stream in west Africa.**

ABOVE **A male blue bareleg *Heliocharis amazona* from Ecuador.**

Euphaeidae (gossamerwings)

These robust, thickset damselflies are largely restricted to India and South-east Asia, although one species occurs in the Middle East and south-eastern Europe. The bodies of most species have sombre colours, but the wings are often brightly tinted with metallic greens or reds, thickly bordered with black. They breed in woodland streams and have a fluttering flight, quickly perching again after they have become airborne. The larvae hide under stones and are unique amongst Odonata in possessing seven pairs of long filamentous gills down each side of the abdomen, in addition to three sac-like caudal lamellae at its tip.

Polythoridae (bannerwings)

This family is restricted to the tropical rainforests of Central and South America. All species are large, and several have enamel-like blue bodies. Their wings are broad, and in many species are iridescent and marked with white, yellow, orange or black patches. The wing markings of some species give them a close resemblance to clearwing butterflies (of the subfamily Ithomiinae) with which they share forest clearings. These butterflies contain toxins that make them distasteful to predatory birds, and mimicking them may help to protect the polythorid damselflies from predators, too.

Polythorids breed in shady forest streams, which may be full of dead leaves. The larvae have six pairs of supplementary gills underneath the abdomen and their caudal lamellae are swollen and equipped with finger-like projections. Adult males are territorial, and many species defend sunny patches. Other species establish small, well-defined territories, which they may defend for up to three weeks.

Synlestidae (sylphs)

These large, slender black or metallic green damselflies live in Africa, Australia and

BELOW **A female European gossamerwing *Epallage fatime* perched near a shady stream in western Turkey.**

BELOW RIGHT **A male bannerwing *Polythore mutata* from the forests of Peru. In flight it resembles a toxic butterfly and so birds avoid it.**

LEFT **A male *Lestoidea barbarae* from Queensland, Australia, where it breeds in forest streams.**

China, with one species also occurring in the Caribbean. Adults characteristically hang down vertically from vegetation when perched, with their wings held open. Most breed in cool forest streams.

Lestidae (reedlings)

Lestid damselflies are similar in general appearance to Synlestidae, although most lestids are smaller. They, too, perch with their wings held open, and most are coloured metallic green; many have blue spots at the base and tip of the abdomen.

This family occurs throughout the world. Many breed in exposed sunny pools that dry out during the summer. In the absence of fish, which would otherwise eat the conspicuous larvae, the large, active, predatory larvae can complete development in the warm waters in just a few weeks. Some lestids that live in north temperate regions are unique amongst Odonata in that they spend the winter hibernating in the adult stage.

Lestoideidae

This is a small family of generally large damselflies. There are three genera, one occurring in eastern Australia and Papua New Guinea, one restricted to north-eastern Australia and the third found in northern India, Myanmar (Burma) and China. They breed in forested streams. The larvae have supplementary abdominal gill tufts, and some species can survive in seasonally dry streams.

Megapodagrionidae (flatwings)

Over 200 species have been described in this family. They live in tropical rainforests throughout the world, although only a few species are known from Africa. Many species have long slender abdomens, and perch with their wings held open in deep shade close to

ABOVE ***Amanipodagrion gilliesi*** **is found in the Congolese forests of west Africa.**

ABOVE RIGHT **The shortwing *Perissolestes castor,* which flits like a ghost through the shady rain forests of Peru.**

forest streams. Their bodies are often black, but sport bright spots of yellow, white or blue on the face and tip of the abdomen that shine like lamps in the forest gloom.

One large species in this family is the giant waterfall damselfly *Thaumatoneura inopinata*, which occurs in Panama and Costa Rica only. Its specialist habitat is waterfalls. Adults characteristically perch on vegetation overhanging the waterfall, drenched in spray. They lay eggs in moss or roots at the side the waterfall, where the larvae will develop.

Perilestidae (shortwings)

Most species in this family live in Central and South America, although one is known from the tropical rainforests of West Africa. Perilestids probably evolved in this region when South America was joined to West Africa 160 million years ago, during the Jurassic period. All species have very long, slender bodies but short wings that only extend about half way along the body. They live near shady forest streams, where they are perfectly camouflaged by the brown and black mottling on their bodies, and by their habit of flying only short distances at an even pace. The larvae live among dead leaves in shallow streams.

Hemiphlebiidae

This family includes just one species, *Hemiphlebia mirabilis*, which is restricted to south-east Australia and Tasmania. Certain features of the wing venation set it apart from all other living species of Odonata, and suggest that it may have the oldest ancestry of all living odonates.

This damselfly lives in damp marshy areas with dense growths of reeds, where it is perfectly camouflaged by its small size and metallic green coloration. At the tip of the abdomen are two pairs of long, brilliant white appendages, which the insect displays by waving its abdomen. The larvae live in shallow water, and the species is able to survive seasonal drought in the egg stage.

Coenagrionidae (pond damselflies)

Coenagrionidae includes more species than any other damselfly family. They breed in standing or flowing water in tropical and temperate regions throughout the world, and the family includes some of the most widespread and abundant of all damselflies. Almost any freshwater habitat in the world will support a species of coenagrionid. Most are blue and black, but others are red. Some, like *Agriocnemis pygmaea*, are among the smallest of all damselflies, measuring only 16 mm (2/3 in) in length. This species typically lives in large swarms in damp grassland and rice fields in South-east Asia.

Species in the genus *Ischnura* undergo long migratory journeys, borne on frontal weather systems, and have colonized even remote Pacific islands. They will also tolerate quite extreme habitats, including temporary pools, saline lakes and hot sulphurous springs. Some coenagrionids have very wide distributions. For example, the marsh bluetail *Ischnura senegalensis* occurs throughout much of Africa and Asia, and *Enallagma cyathigerum* occurs in North America, Europe and Asia.

While the larvae of most species are aquatic, those of a few species of *Megalagrion*, a genus that is found only on the islands of Hawaii, are terrestrial and live among dense forest leaf litter. Some of these unusual damselflies also breed in water-filled rot-holes in forest trees or in the water that collects in the leaf bases of bromeliads, plants that grow on the trunks of huge trees in tropical rainforests.

Isostictidae (narrow-wings)

This family is confined to Australia. All species have very long, slender bodies that are unobtrusively coloured pale blue, grey and black. They live in shady forests, frequenting dense shade and overhanging vegetation. These features, together with their slow flight, makes them very difficult to see and gives them an almost ghostly appearance. They breed in forest streams and waterfalls.

LEFT **The large red damselfly *Pyrrhosoma nymphula* is common in ponds throughout most of Europe, where the larvae live among leaf debris at the bottom.**

Platycnemididae (brook damselflies)

Most species in this Old World family breed in flowing water. They occur throughout temperate and tropical regions. Males in this family are distinguished by their flattened tibiae, which may be bright blue or white, and are used in signalling to mates and rivals. Most species are black, spotted with blue or yellow, but a few are entirely red, pale blue or white.

Platystictidae (forest damselflies)

These damselflies are restricted to tropical forests in Central and South America and in Asia. They have long slender bodies that are typically black, with a white or pale blue spot at the tip of the abdomen. When one of these damselflies flies in a shady forest, this spot is often all that is visible as a tiny fleck of light floating in the gloom. The front of the face or thorax is also often marked with a bright stripe of blue or pale yellow, and males may confront each other at close range, hovering face-to-face. Females lay their eggs in plants overhanging the forest streams in which the larvae later develop.

BELOW **The broad, flat legs characteristic of male Platycnemididae are visible in this white-legged damselfly *Platycnemis pennipes*. This species occurs throughout most of Europe and breeds in slow-flowing rivers and sometimes ponds.**

BOTTOM **A male forest damselfly *Palaemnema desiderata* in Belize. Eggs are laid into woody plants overhanging streams.**

Protoneuridae (pinflies)

Like platystictids, this family of small, slender damselflies is also found in tropical forests. Distributed worldwide, they also favour shady streams, though many breed by preference in pools or slow-flowing stretches of streams. Most are dark-bodied, but with bright patches of blue, red or orange on the head, thorax and tip of the abdomen, which gleam in the pools of sunlight that penetrate the dense forest canopy. They fly close to the surface of the water, and many species lay their eggs in dead, rotting wood protruding from the surface.

Pseudostigmatidae (forest giants)

All the species in this family are the largest of the world's damselflies. In fact, *Megaloprepus caerulatus*, the helicopter damselfly of Costa Rica, has, at 18 cm (7 in), the largest wingspan of any odonate. Pseudostigmatids live in the tropical rainforests of Central and South America. They all have long narrow wings and enormously elongated abdomens that may reach 21 cm (8 1/4 in) in length. Most species have drab body colours, but the wings

FAR LEFT **A male pinfly *Psaironeura tenuissima* from the hot humid forests of Ecuador. The bright orange patches glow like beacons in the forest gloom.**

LEFT **This female forest giant *Mecistogaster rotundata* from Peru lays its eggs in water-filled hollows in fallen trees. It has a wing span of 132 mm (5¼ in).**

of some are marked white or yellow at the tip, and *M. caerulatus* has a broad black stripe across the middle of each wing that shimmers metallic purple in the sunlight.

Adults are usually encountered in forest clearings away from obvious signs of water, hovering over fallen trees. They specialize in feeding on orb-web spiders. As they hover in front of the web, the vibrations created by their wingbeats cause the spiders to come out onto the web to investigate, whereupon the giant damselfly deftly plucks the spider from the web, backs away to avoid becoming entangled in the web, and eats its catch.

The long bodies of the adults enable them to lay eggs into the water that gathers in rot-holes in trees either by inserting the eggs directly into the rotten wood or by flicking the eggs into the water. The larvae feed on mosquito larvae, which breed in profusion in this habitat, but at times of food shortage they will resort to cannibalism.

Suborder Anisozygoptera

Epiophlebiidae

This family includes only two living species, and these are the sole living representatives of the entire suborder Anisozygoptera, which, however, has a diverse fossil record. In appearance, the body is similar to dragonflies of the family Gomphidae (see below) but the wing venation is similar in several features to that of damselflies. The larvae, however, look just like dragonfly larvae.

One species, *Epiophlebia superstes*, occurs in Japan, the other, *E. laidlawi*, in the eastern Himalayas. Both species breed in fast-flowing, cold mountain streams and may take from five to nine years to complete development. In both species adults are on the wing in cool temperatures early in the season. *Epiophlebia laidlawi* occurs at altitudes of 3000-3700 m (9800 ft), and breeds in waterfalls above 2000 m (6500 ft). In both species, the females lay their eggs endophytically.

Suborder Anisoptera (Dragonflies)

Aeshnidae (hawkers)

This family has a worldwide distribution. Aeshnids are big dragonflies with huge eyes that more or less meet in a broad line over the top of the head. They are seemingly tireless fliers, and are seldom seen to perch. Their bodies are usually black or dark brown, mottled with small spots of bright blue, green or yellow. In some species the abdomen may be largely red, brown or green. Most species breed in open pools but there are also forest species that breed in running water. The larvae are large and streamlined.

This family includes some of the world's largest dragonflies. For example, *Tetracanthagyna plagiata*, which has a wingspan of 160 mm (6 1/4 in) and the giant emperor *Aeshna tristis* with a body length of 120 mm (4 3/4 in). Most adults are on the wing during the heat of the day, but some species will continue to fly into the twilight hours: the majority of species in the tribe Gynacanthini fly between dusk and dawn, and are characterized by their very large eyes and rather drab brown and green body colours. During the day, they roost in dense thickets.

Some aeshnids routinely migrate over vast distances. For example, the vagrant emperor *Hemianax ephippiger* breeds in temporary pools in arid areas of North Africa. Of necessity newly emerged adults must move on to find suitable new breeding habitats. Sometimes they wander far afield and turn up in northern Europe and occasionally in Iceland. The green darner *Anax junius* regularly commutes up and down the eastern seaboard of North America, but that species, too, sometimes gets blown off-course, and has even been known to turn up in southern England.

TOP **A male hawker *Indaeschna grubaueri* in a Malaysian forest.**

ABOVE **A male clubtail *Gomphus pulchellus* from France, absorbing heat radiating from a cow-pat.**

Gomphidae (clubtails)

Adults of this family are distinguished by the swollen tip of the abdomen and by their smallish eyes, which do not meet over the top of the head. Most species are black, with stripes and spots of yellow, green or pale blue. Almost all breed in flowing water. They occur throughout the world in temperate and tropical regions.

Females lay their eggs exophytically. The larvae are typical bottom-dwelling 'sprawlers', with a broad abdomen and splayed legs, but a few also live buried beneath the substrate and have long breathing siphons at the tip of the abdomen. The larvae of some species are limpet-shaped, which helps them cling to stones in torrential streams. Larval development can be protracted, in some cases taking up to five years.

Most gomphid species have a relatively short flying season. In the tropics, the adults emerge shortly after the start of the rains. In temperate regions, they emerge in spring or early summer. Adults patrol territories, comprising short stretches of stream bank or patches of streams in forests where the sun is able to penetrate. They frequently rest on vegetation or boulders in the watercourse. They quickly disappear into the canopy when the sun goes behind clouds.

Neopetaliidae (redspots)

This family includes just nine species: three are known from Chile, the rest from south-east Australia. This suggests that the Neopetaliidae evolved before the break-up of the supercontinent Gondwanaland (which included all of the continents currently surrounding the Pacific Ocean plus Antarctica), about 160 million years ago. All the species are characterized by a row of red spots along the leading edge of each wing. All occur in cool, moist, mountainous regions and can fly at air temperatures as low as 9°C (48°F). Eggs are laid in moss at the edge of fast-flowing streams. The larvae of some species leave the water at night to hunt on land.

Petaluridae (petaltails)

This is one of the most ancient families of dragonfly, with fossils known from the Jurassic period (195–136 million years ago). Species occur right round the Pacific Basin, in Japan, Australia, New Zealand, Chile, and western North America. They are large dragonflies, with a wing span up to 160 mm (6 1/4 in), well-separated eyes and large petal-shaped appendages at the tip of the abdomen. They are darkly coloured with bands of pale yellow, which helps to camouflage them against the tree trunks on which they frequently settle. They have a powerful flight, but are not particularly agile, and are easily caught by predatory birds.

BELOW **A male grey petaltail *Tachopteryx thoreyi* in Florida, USA, concealed against the bark of a tree.**

Petalurids live in bogs and woodland, and breed in shallow seepages and streams. The females lay their eggs into moss. The larvae of some species construct a burrow, in which they spend the day submerged in the water that collects at the bottom, although the entrance to the burrow may be some distance from the stream edge. At night, they leave the burrow to hunt on land. The larvae of other species live beneath damp leaves.

Cordulegastridae (goldenrings)

These large dragonflies characteristically have black abdomens ringed with gold. The females have a long blade-like extension of the ovipositor that protrudes conspicuously beyond the tip of the abdomen. Hovering above shallow streams, they use this to push their eggs deep into the gravel and sand. The larvae are hairy, and lie concealed beneath a covering of silt and debris. In the cool mountain streams where many of them live, they can take many years to develop. In the Himalayas, they occur at altitudes up to 4100 m (13,450 ft). Adult males of most species patrol fast and low down long stretches of stream before turning to retrace their path. Although they can be fast and active they often settle, hanging by their legs from low plants overhanging the stream. The family is found throughout North America and in most of Central America, as well as in Asia, Europe and North Africa.

Chlorogomphidae (tiger bodies)

Species in this family superficially resemble members of the Cordulegastridae. However, unlike those of the cordulegastrids, the wings of females, and some males, are marked with large black patches bordered with white or yellow, and the females lack the long oviposition blade. These dragonflies are confined to mountainous regions of India, Japan and South-east Asia, where the larvae burrow into the sand at the bottom of pools in torrential streams. The adults are among the largest of all dragonflies, their broad wings having a span of up to 145 mm (almost 5¾ in). They live in dense forests, gliding effortlessly along mountain streams, or soaring high above the tree canopy.

BELOW **Male golden-ringed dragonflies *Cordulegaster boltonii* are a common sight patrolling upland streams throughout much of Europe.**

LEFT **A female river emerald *Macromia georgina* from Florida, USA.**

Macromiidae (river emeralds)

Most species in this family have metallic green bodies with yellow or orange spots. Both the larvae and adults have long, spidery legs. They occur throughout the world, the larvae living among plants in lakes and rivers, while overhead the fast-flying males rarely settle. Females lay their eggs alone, skimming the water while tapping the tip of the abdomen onto the surface to wash off the eggs.

LEFT **A male southern emerald *Eusynthemis nigra* from rainforest on Mt Lewis, Queensland, Australia.**

Synthemistidae (southern emeralds)

These small dragonflies have a narrow green abdomen marked with yellow spots, and a rather hairy thorax. They occur in Australia, New Guinea and New Caledonia, where the squat, hairy larvae of most species live buried in muddy seepages or slow-flowing rivers. The eggs and larvae of some Australian species are able to withstand periods of drought.

Corduliidae (emeralds)

This large family of dragonflies occurs throughout the world. Most are metallic green or black, with bright green eyes, and often with yellow or orange spots on the abdomen. Many species are confined to forests and woodlands, where males are commonly seen patrolling the edges of lakes and rivers fringed with trees, where they frequently pause to hover and inspect inlets for ovipositing females. The genus *Somatochlora* includes several species that live in far northerly latitudes, including Sahlberg's Emerald *S. sahlbergi*, which breeds north of the Arctic Circle and has the most northerly distribution of any known odonate. The larvae are broad-bodied and spider-like, and live concealed beneath debris and leaf-litter.

Libellulidae (chasers, skimmers, darters)

This family includes over 1000 species – more than any other family of Odonata. Libellulids occur throughout the world, and breed in every habitat suitable for odonates. Some species are confined to tropical forest streams,

BELOW **The brilliant emerald *Somatochlora metallica* breeds in tree-lined ponds throughout much of northern Europe.**

LEFT **A male dropwing *Trithemis persephone*, a member of the large family Libellulidae, basks on a sunny rock in Madagascar.**

but others travel the world, breeding in temporary pools. However, libellulids are most conspicuous around sunny, unshaded lakes and ponds, where many different species may gather in their hundreds. Males characteristically perch upright on prominent branches, twigs and plants emerging from the water, from which they take off to pursue prey, mates or rivals before returning to the same perch.

Most species have red, blue or yellow bodies, but a few have metallic colours. Many have patches of yellow or brown on the wings. The thorax and abdomen of the males (and some females) of several species become covered with a dusting of grey, white or blue wax (called pruinescence) as they mature.

Most are medium-sized dragonflies, with bodies roughly the same length as the wingspan. The larvae of most species are squat and hairy, and live buried in debris and silt.

BELOW **This libellulid, *Zenithoptera americana*, has opaque wings that protect its thorax in hot weather.**

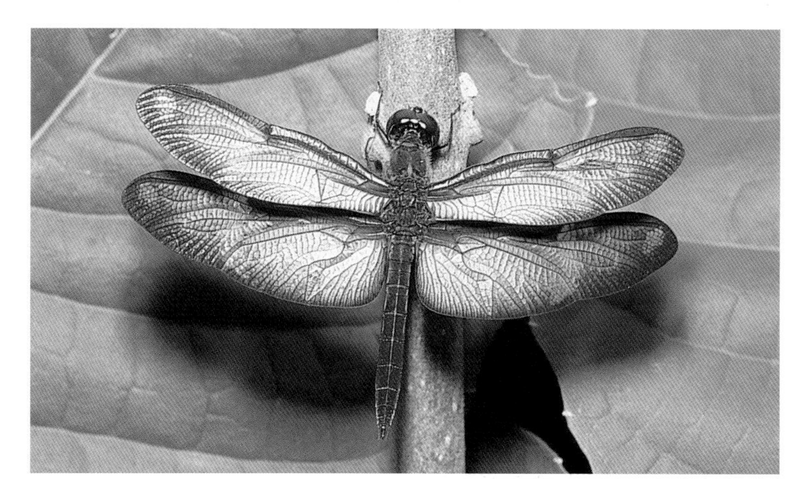

Odonata and humans

Odonata have inspired humans for millennia. But, increasingly, the influence of humans on Odonata has not been benevolent as habitat destruction has taken its toll. Nevertheless, Odonata respond rapidly to positive conservation measures. They are easily attracted into the garden by digging a pond where they can be studied in detail.

Odonata in folklore

Dragonflies and damselflies have long impinged on the human psyche throughout the world. There are many writings and images featuring these insects from Europe and Asia, some over two thousand years old. Although some written accounts focus on the more sinister aspects of odonates, many refer to their aesthetic appeal, their beauty and grace.

In Japanese culture, odonates are revered as symbols of courage, strength, victory and happiness, and they sometimes have a religious significance, too, being associated with the visitation of spirits to the home. Indeed, an old Japanese name for Japan, *Akitsu shima*, bestowed on the country by its first emperor, means 'Island of Dragonflies'. Odonata are frequently the focus of haiku, a stylized form of Japanese poetry (see below).

In European tradition, on the other hand, odonates are generally considered more threatening: indeed, they are often seen as associates of the devil. This is reflected in some of the traditional English names for odonates, including 'devil's darning needles' and 'devil's riding horses'. However, this perceived relationship has not always been the case, since in Norse mythology Odonata were linked with Freya, the goddess of love, wealth and eroticism. The link between odonates and snakes has ancient roots in

Haiku

Haiku is an ancient form of poetry practised for centuries in China and Japan, but thought to have originated in Persia. Haiku is characterized by its simplicity and brevity: the words are usually arranged in three lines of 5-7-5 syllables, and used to express one or two clear interrelated images. Abstract thoughts are avoided; instead the words provide a mind-picture which the reader explores. The mind-images must be placed in their correct seasonal context – Odonata are associated with summer and autumn. Here are a couple of examples of haiku that feature dragonflies:

Crimson pepper pod
Add two pairs of wings, and look
Darting dragonfly

The beginning of autumn
Decided
By the red dragonfly

ABOVE **The female broad-bodied chaser *Libellula depressa* resembles a large wasp and can cause people to be fearful of this species.**

European culture, dating back 3000 years, and can be seen in vernacular names such as 'adder's needle', 'ox-viper', 'adder's servant' or 'adderspear'. The widely used name 'dragonfly' also refers to this association with serpents and the devil.

Traditionally, Odonata are thought to possess a fearsome sting – leading to names such as 'horse-stinger' and 'bullstang'. This supposed ability is still widely believed and the presence of odonates can provoke fear in some people, particularly when confronted by dragonfly species resembling very large wasps. The sudden arrival of migrating swarms of dragonflies can provoke mass panic, and a recent newspaper article reported on the abandonment of a merchant ship by its crew off the Philippines as a result of such an event.

Odonata as valuable assets

Magical and medicinal properties have been assigned to odonates in China and Japan. They are said to cure syphilis and promote sexual prowess, and are used as a treatment for asthma, fever, tonsillitis and eye conditions. The eggs of Odonata were used in ancient Mesopotamia to treat menstrual pains. Adult and larval odonates form part of the diet of many people in Asia, Africa and Central America. Usually, they are cooked before being consumed, but if they are eaten raw people risk infection by parasitic worms (flukes) that live within the larvae, using them as intermediate hosts.

Odonata also have value as predators of insects that are harmful pests of humans. The most significant of these pests are mosquitoes, the adults of which transmit malaria. Malaria is responsible for the deaths of millions of people throughout the world every year.

Mosquito larvae are prevalent in rice fields, but predation by odonate larvae can account for a 90-100% reduction in their numbers. In places where piped water is not available, people often store drinking water in containers in their homes. These are rapidly colonized by mosquitoes. However, the addition of just one or two dragonfly larvae to the container can control this potential source of disease, and – as long as these larvae are not accidentally consumed – there is no risk of parasitic infection of people who drink the water. Adult Odonata also account for large numbers of adult mosquitoes.

Other major insect pests whose numbers may be controlled as a result of predation by adult Odonata include sandflies (which infect humans with the disfiguring disease leishmaniasis), locusts, wood-boring beetles, and moths whose larvae consume rice and cotton.

Because of the complex and varied demands on the ecosystem made by each stage in their lifecycle, Odonata are valuable indicators of habitat quality. The larvae of many species are sensitive to organic pollution, which may result from malfunctioning sewage works or the run-off of fertilizers from agricultural land. This can lead to a reduction of oxygen in rivers and the mass deaths of fish as a result. Insecticides may also be washed into watercourses from adjacent fields, and these, too, can cause serious damage to ecosystems. These toxins are frequently passed up the food-chain via odonate and other insect larvae to fish, and eventually to humans. For these reasons, a reduction in the abundance and diversity of odonates is often an indication of a decline in water quality.

Odonata can also be useful in assessing the conservation value of a site. They are sensitive to changes in habitat structure, which may be brought about by clearance of aquatic plants; straightening, dredging and canalization of rivers; clearance of bankside vegetation; and clearance or selective logging of forests. Because they are near the top of the food-chain, odonates act as an early warning system of significant declines in animals at lower levels of the food chain, involving far greater numbers of species.

Another reason for the value of Odonata in assessing conservation value is the ease with which the adults can be surveyed. They are easily seen, and concentrated near water bodies, so a survey representative of the whole fauna can quickly be achieved. In addition, there are relatively few odonate species when compared with other groups of insects, and the taxonomy is well resolved, so there are few identification problems, even in tropical regions.

Human impacts on Odonata

Odonates are the most ancient group of flying insects, having lived on this planet for almost 300 million years. They are a very successful group, and are supremely adapted to their style of life. However, along with most other groups of animals and plants, they currently face one of their most taxing periods for survival. In the relatively short time that humans have inhabited the earth, and especially in modern times, there has been an unprecedented rate of species extinction – including odonates – as a direct result of human impacts on the environment.

The known human impact on odonate faunas has probably been most severe in developed regions where intensification of farming has lead to the widespread destruction of wetlands. Huge areas of bogs, fens and marshland have been drained and converted to arable farmland. Ponds have been filled in, and rivers and streams canalized, piped underground or turned into open sewers or lifeless drains, because of the impact of fertilizers and pesticides. In many countries, this has resulted in the local extinction of odonate species; a contraction in the range of many species, leaving them more vulnerable to extinction; and a reduction in the abundance of many species that were once considered common.

The diversity of odonates is greatest in tropical forests, and the rainforests of northern South America, West Africa and the Philippines support large numbers of species that are endemic to each region (that is, found

Casualties of modern agriculture

Populations of Odonata are declining throughout the world due to degradation of their habitats by human activities. Wetlands have been damaged or destroyed by over-abstraction of water, by run-off of pesticides from field-sprays and sheep-dips, and by run-off of fertilizers causing nutrient enrichment that promotes the growth of algae and toxic cyanobacteria and results in deoxygenation of the water. In Brazil, for example, one of the hawker dragonflies, *Aeshna eduardoi*, and a sprite dragonfly, *Leptagrion siquieri*, are threatened by deforestation, whereas it is afforestation that threatens the two-toothed golden-ringed dragonfly *Cordulegaster bidentatus* in Germany.

In Britain, drainage of wetlands as a result of agricultural intensification has resulted in a loss of 75% of ponds since 1888, while over 90% of peat bogs have been lost to extraction, agriculture and forestry industries. Pollution of the West Moors river in southern England is thought to have been the cause of extinction from Britain of the orange-spotted emerald dragonfly *Oxygastra curtisii* in 1963.

The endemic Hawaiian damselfly *Megalagrion pacificum* is threatened by the introduction of insectivorous fish to combat mosquitoes. In New Zealand, two species in the damselfly genus *Xanthocnemis*, which until recently lived in separate habitats, have begun to interbreed, threatening the existence of one of the species. Destruction of upland forests has allowed the lowland species to move into the mountains and meet the other species, which formerly was isolated in upland forests.

nowhere else in the world), and localized in relatively small areas. However, human pressure on many of these forests is enormous and large swathes are being destroyed every year. The odonatan faunas of these forests are still only poorly known, so the impact of deforestation can only be guessed at – although it is likely to be catastrophic.

Human impacts on the environment are not always deleterious to odonates. Clearance of pockets of forest and secondary forest regrowth can lead to an increase in the number of dragonfly and damselfly species in a locality. This is because – provided they are big enough – the patches of primary forest allowed to remain still support many of the original forest species, while opening of the canopy and the increase in extent of forest-edge habitats allows other species associated with open habitats to colonize the locality. However, although there may be a net increase in species following small-scale logging, this may be at the expense of some localized forest species, which in the tropics may be endemic. Furthermore, many of the new colonists are likely to be species that are common and widespread elsewhere because they are tolerant of degraded habitats.

RIGHT **Small-scale peat cuttings like this one in Northern Ireland can create superb odonate habitats.**

Odonates may benefit from small-scale cutting of peat bogs, since this creates small pools and leaves the water quality of the bog intact. However, large-scale industrialized peat cutting utterly destroys such habitats. Emission of sulphur and nitrogen from industrial processes and car exhausts leads to the deposition of acid rain and the acidification of lakes and streams, resulting in declines in many animal species. However, some species of Odonata have benefited from such acidification, because of the elimination of predatory fish and other competing species of odonate.

In the last ten years or so, many species of European odonates – such as the small red-eyed damselfly *Erythromma viridulum* and the migrant hawker dragonfly *Aeshna mixta* – have significantly extended their ranges northwards. This is likely to be in response to global warming caused by increases in atmospheric carbon dioxide and methane resulting chiefly from air and marine pollution, deforestation and agricultural intensification. If global warming continues, as is forecast, then we can also expect to see a contraction in the ranges of odonate species restricted to northern latitudes and high altitudes.

Conservation of Odonata

In order to halt the decline in the abundance and diversity of Odonata, it is essential to protect the habitats in which they live. Efforts to protect individual species threatened with extinction, by legislation that makes it illegal to exploit or kill individuals, are futile if their habitats are not protected from degradation or destruction. In any case, the populations of most species are robust enough to tolerate the removal of a few specimens.

Wildlife conservation is not a new concept. People have protected forests and the animals they support for centuries, but until recently this was done only so that certain animals could be hunted and exploited. Many other non-target species incidentally benefited from the setting aside of hunting reserves. It is only in the last century that a conservation movement has sprung up with the goal of protecting species in their own right.

One of the world's oldest nature reserves was established at Wicken Fen in Cambridgeshire, eastern England in 1899. British naturalists at the time were alarmed at the speed with which the East Anglian fens had been drained and converted to arable farmland, and wanted to protect Wicken Fen, which they valued as an excellent location for collecting moths and butterflies. Since then, a worldwide nature conservation movement has sprung up that now includes millions of active members.

These conservationists have focused most of their endeavours on large animals, birds and plants but insects, including dragonflies and damselflies, have also benefited from the wildlife reserves that have been established. For example, in Britain all but one species of odonate occurs on a nature reserve, although none of the reserves have been specifically set up to conserve odonates. In Japan, on the other hand, over 20 reserves have been created especially for these insects.

In recent years, Odonata have gained in popularity among scientists who study their behaviour and taxonomy, and among amateur naturalists who are interested in mapping their distribution or simply enjoy watching and photographing them. The results of this increase in interest has been a greater understanding of the habitat requirements of many species, informed opinion about how best to conserve them, and the organization and coordination of conservation effort. This has led to the establishment of an international dragonfly society, as well as many national ones, which in turn has further promoted interest in Odonata. Today, odonates are the focus of many local, national and international conservation programmes.

Recording Odonata

The most fundamental task in conserving any group of organism is to establish the distribution and relative abundance of that organism. It is important to determine which species are restricted regionally because of climatic constraints, and which are found only in particular habitats – and where those habitats occur. The combined effort of many enthusiastic amateur naturalists is essential to this task of mapping.

There are active Odonata Societies in many European countries, including Germany, France, The Netherlands and Britain, as well as enthusiastic individuals in other European countries; consequently the distribution and conservation status of most European species is quite well known. In the USA and Japan there are national Odonata Societies engaged in active recording and collecting to build up a picture of the distribution and ecology of the native Odonata species. In other countries around the world including, South Africa, Australia, India, Thailand and Malaysia, there are small groups of individuals working on the conservation of Odonata.

In Britain, the Odonata Mapping Scheme is currently coordinated on a voluntary basis by the British Dragonfly Society, with support from the Government-funded Biological Records Centre. During the past 30 years, over 160,000 records have been received from more than 2000 amateur contributors. A fairly even coverage of about 87% of the country has been achieved, and now the current status and habitat requirements of all the resident and migrant species of Odonata that occur in Britain is known. As a result, certain locally and nationally threatened species have been targeted for special conservation effort.

The proportion of Odonata enthusiasts to species of Odonata in Britain is high, certainly higher than in any other country in the world. In tropical countries, where the diversity of species is the highest, the number of such enthusiasts is the lowest. This poses a great challenge for the future worldwide conservation of these insects.

Legislation

Practical conservation of Odonata can be achieved using three principal measures: the legislative protection of species; the conservation of habitats; and the creation of new habitats. Legal protection of insect species is usually ineffective, and may actually be counter-productive. Most insects produce large numbers of offspring, and even rare species with restricted distributions may have relatively large populations that can withstand the loss of small numbers of individuals. The exception may be some of the larger dragonflies, which require such big territories that only a few individuals could be supported at a single site. In this case, removing even a few individuals might have a serious impact on the population.

Some countries have adopted blanket protection of all odonate species. For example, in Germany it is currently illegal to kill or even capture any odonate. However, rather than being an effective conservation measure, such apparently laudable legislation brings with it a risk of stifling interest in research by making it difficult to study the insects and compile accurate distribution maps.

Protecting habitats

A more effective measure is to protect odonate habitats, especially rare habitats, habitats of particular regional significance, or those that support large and diverse assemblages of Odonata. However, it is essential that these reserves remain inviolate. All too often, governments designate areas as nature reserves, only to allow damaging activities on them in the name of national or economic interest. As long as nature reserves are allocated minimal monetary value in cost-benefit analyses, they will always be vulnerable to development.

Designation of areas as nature reserves is not always enough to protect the species that live in them. The value of a site may decline through neglect. This is especially a problem at nature reserves that are too small to allow a natural turnover of different habitats, or on sites including habitats that have resulted only from the activities of humans. Continued active management is often essential to maintain the nature conservation interest of a site. For example, many heathlands are extremely valuable habitats for odonates in Europe, but owe their existence to the activities of humans who have cleared woodland from poor, sandy soils in order to graze livestock. In the absence of grazing animals, the woodland quickly re-establishes itself, and the value of the site for dragonflies and damselflies is diminished.

Creating habitats

Habitat creation can be another effective means of conserving or enhancing populations of Odonata. Because of their excellent powers of dispersal, many species are adept at colonizing new sites. With good planning, an artificially created site can offer a suitable mosaic of different habitats, with various depths of water, sufficient submerged, emergent and bankside plants, so that it can quickly support a diverse and abundant odonate fauna.

New sites created near existing odonate populations are likely to be most successful, but even sites remote from known populations will be colonized. Most of the early colonists will be species (called eurytopic species)with broad habitat requirements and these are generally widespread and common. However, as the site matures and enters later stages of vegetational succession, less common species may become established. At such a site created at Ashton Wold in Northamptonshire, central England, careful management, specifically targeted at optimizing habitat features to support a diversity of Odonata, has resulted in the number of species breeding on the reserve increasing from two to fifteen in less than ten years.

ABOVE **The 'Dragonfly Kingdom' at Nakamura, Japan, was created specifically for odonates and attracts many human visitors as well.**

Creating a pond for Odonata

Anybody, whether they live in a town or in the countryside, and no matter how big or small their garden or backyard, can live close to odonates by creating a suitable pond. The size of the pond is not crucial, although big ponds with long margins will attract more species and greater numbers of individuals than small ones. By following a few simple guidelines, you can soon persuade these beautiful and fascinating insects to live close to your home.

The pond should be of varying depths. At its deepest, it should be at least 1 m (about 3 ft) deep, to provide a refuge that will not freeze during the winter. The profile should gradually shelve towards the margins, to provide an extensive shallow-water zone that occasionally becomes exposed during the summer, and a swampy margin that experiences periodic flooding. You can use a butyl liner to hold the water, though for larger ponds, a clay lining is more appropriate.

Ideally, you should allow your newly created pond to fill with rainwater. The presence of even small quantities of phosphates in tap water will encourage the growth of planktonic algae, which give the water a green and turbid appearance, or unsightly clumps of filamentous algae. However, provided you do not repeatedly top up the pond with tap water and you provide plenty of aquatic plants, the planktonic algae will soon disappear. Filamentous algae may take a few years before they finally die out, but they do not harm aquatic animals; in fact, they can provide a useful refuge, although there is a risk that they may out-compete and smother some of the larger plants.

Plantlife

You should plant your pond with aquatic plants of native rather than exotic species. Many ponds in the countryside have been choked by the thoughtless release of exotic water-plants. Choose plants that will provide extensive beds of submerged, floating-leaved and emergent vegetation. These will provide refuges and emergence supports for odonate larvae, perches for territorial males and oviposition sites for females. Aquatic plants can usually be bought at local garden centres: do not collect them from the wild. You may find it necessary to remove some aquatic

RIGHT **The rich variety of plants at the Wake Valley Pond in Epping Forest make it an ideal habitat for dragonflies. The pond is less than 16 km (10 miles) from the centre of London but supports a diverse and abundant odonate fauna and is an excellent place to study them.**

plants periodically, to prevent the pond becoming totally choked with vegetation and drying out. However, take care not to remove all the plants in a particular microhabitat at one go, to ensure continuation of all habitat types. You should carefully wash the plants you have removed from the water in the pond or in a bucket so that any animals trapped in the plants can escape. You can also place the plants at the side of the pond for a few days, to allow any remaining trapped creatures the opportunity of returning to the pond.

The pond should be sheltered from the prevailing winds by provision of a tall bank, tall trees or shrubs. However, the water should be exposed to full sunlight, and not shaded by trees. Adding bare soil and logs, some partly submerged, will provide basking sites for adults and additional oviposition sites. A pond exposed to crosswinds or one that is shaded will not be attractive to the adults of most species of odonates. Also, leaves falling into the pond from deciduous trees during autumn can result in deoxygenation of the water as they decompose, and this may kill the odonate larvae. Nevertheless, some leaves, detritus and silt at the bottom of the pond are desirable, because they provide a habitat for burrowing odonate larvae – and their prey. The close proximity of trees, shrubs and tall vegetation will also be attractive to adult Odonata, which will use them as roosting sites, mating perches and feeding sites.

You should avoid stocking your pond with fish (especially carp) or waterfowl. Many fish and waterfowl will eat odonate larvae, although the larvae may be able to hide if there are enough plants in the pond. Carp are a particular problem, because they eat and uproot aquatic plants. Lack of large plants encourages blooms of planktonic algae, which makes the water cloudy, reducing the amount of light that can penetrate the water and impairing the growing conditions for large aquatic plants. Soon, the concentration of oxygen dissolved in the water also declines, and the pond becomes unsuitable for breeding populations of odonates. Waterfowl have a similar effect by eating and uprooting plants, and by adding their nutrient-rich faeces to the water.

Studying Odonata

The first step in studying odonates is to be able to identify them correctly. Fortunately, there is nowadays a wealth of superb field guides to both adult and larval stages of the odonate fauna of many parts of the world. Some of these books are listed in the bibliography. There are also many websites that are worth a visit (page 96).

With a little skill, practice and patience, you will soon learn to recognize the adults of most species when they are perched or even in flight. A pair of close-focusing binoculars is very useful for this. However, it is sometimes necessary to confirm identification by examining the specimen in the hand. Therefore, a good butterfly net is essential. It should have a lightweight, 1–3 m (3¼–10 ft) long, extendable pole and a deep net bag. Dragonflies can be difficult to catch. They fly fast and are very wary of anyone approaching. You should therefore avoid any sudden movements, and try not to cast a shadow over your quarry.

On no account should you capture freshly emerged teneral odonates. Their bodies and wings are very soft and easily damaged. It is best to remove odonates from the net by gently but firmly grasping the outer pair of wings at the base and carefully folding both pairs together over the back of the thorax. The claws of the odonate will be entangled in the net bag, and you should gently release them to avoid damaging them. When you are able to view these splendid insects close up, you will appreciate the beauty of their colours and their complex structure, as well as becoming familiar with the intricacies of identification.

Recording and mapping the distribution of Odonata in your neighbourhood can be very rewarding. As with any other group of animals, it is impossible to conserve odonates without having a good idea of their distribution, which species are rare and which common, and the type and location of habitats that are important for them. Information on breeding is especially important, so make a note of any breeding activity you see, such as the presence of exuviae, pairs copulating, ovipositing or in tandem. Make sure you visit sites at the time of day and in weather conditions that are most likely to encourage adult activity. You can also monitor populations over a number of years, by slowly walking a regular route several times during the year in good weather conditions and counting the number of adults you see of each species. During the winter, you can dredge larvae from leaf-litter or aquatic plants, using a pond net. Make sure the net has a sturdy pole and a strong connection between the net frame and pole, to prevent it from bending.

If you have a pond in your garden, it will provide an ideal place to study the behaviour of all the stages of Odonata. You can carefully mark adults with unique spots of coloured paint on the wings so you can follow the activities of individuals over a number of days or weeks. You can monitor the number of adults emerging from the pond each year by collecting and counting the larval exuviae.

You can also collect the larval stages from your pond and keep them where they are most easily watched, in an aquarium. This will give you the opportunity to study the behaviour and development of odonate larvae and also witness the emergence of the adults. Fill the aquarium with water from your pond, and add gravel and mud and plant debris from the bottom of the pond. You should also introduce a few aquatic plants. Be sure to add a few sticks that stand 10–20 cm (4–8 in) above the surface of the water, to provide the larvae with emergence supports.

It is possible to keep several damselfly larvae together, but you should keep dragonfly larvae separately or they will eat each other. They can be fed with small insects and waterfleas that you can collect from your pond; alternatively, you may be able to buy live waterfleas or bloodworms from an aquarist. You can feed large dragonfly larvae on small earthworms. Add a few water snails (limnaeids with pointed shells obtainable from aquarists) to consume any dead animal matter. Don't forget that adult and larval odonates make superb subjects for macrophotography – as can be seen from the photographs in this book.

Glossary

aestivation suspension of development during the summer

bursa part of the female genitalia: a sac for sperm storage

caudal relating to the rear of an animal

caudal lamellae the three conspicuous leaf-like organs situated at the rear of the abdomen in damselflies and used as respiratory and swimming organs

chitin a protein-like substance, the main constituent of an insect's cuticle

conspecific individuals belonging to the same species

crepuscular species that are active at dusk (such species are also often active at dawn and during daytime in dense forest)

cuticle tough, waterproof, protective outer layer covering the body of insects and many other members of the great group of arthropod invertebrates

diapause a resting stage during which development is suspended, in anticipation of conditions unfavourable for uninterrupted development

endophytic oviposition the act of laying eggs within the tissues of plants

epiphytic oviposition the act of laying eggs onto the surface of plants

exophytic oviposition the act of laying eggs without placing them in contact with plants

exuvia (*plural:* **exuviae**) the shed larval skin

hamules a pair of hooked structures in the male secondary genitalia that he uses to locate and grip the female genitalia during mating

hibernation suspension of development during the winter

labium the 'lower lip', which in odonate larvae is highly modified to catch prey

lateral relating to the side of an animal

ocellus (*plural:* **ocelli**) a simple eye made up of a single lens, one of a group of three such eyes located on the top of the head

ommatidium (*plural:* **ommatidia**) one of thousands of individual facets that make up the compound eye

oviposition the act of egg-laying

ovipositor structure below the tip of the female abdomen used to lay eggs

pronotum the upper surface of the prothorax

prolarva the first larval stadium (stage); it differs from later larval stadia in having no functional legs or mouthparts

prothorax the first thoracic segment, bearing the forelegs, situated between the head and synthorax

Protodonata an ancient group of insects from the Carboniferous period (about 350 million years ago), thought to be ancestral to the Odonata

pruinescence a pale blue or white waxy bloom that develops on the surface of the abdomen and thorax of some dragonfly species as they mature (especially in males)

pterostigma a pigmented area on the leading edge of each wing near the wingtip

secondary genitalia structures used in copulation that are situated beneath and within the second and third segments of the male abdomen.

seta (*plural:* **setae**) hair-like structures on the body surface

siccatation suspension of development during the dry season

spermatheca part of the female genitalia: a sac for sperm storage.

spiracles small openings in an insect's body wall, mostly situated on either side of each thoracic and abdominal segment, that allow air to enter the tracheal system

stadium (*plural:* **stadia**) a stage in the development of a larva; each time a larva casts its skin it enters another stadium, and a larva may pass through 8-18 stadia, depending on the species. The prolarva is the first stadium.

synthorax the fused mesothorax and metathorax, bearing the midlegs and hindlegs and both pairs of wings

systematics the science of classifying the living world according to the presumed evolutionary relationships between different organisms

teneral a freshly moulted larva or adult in which the cuticle has not yet hardened: teneral adults are sexually immature and are characterized by their shining wing membranes and weakly pigmented bodies.

trachea air-filled tubes forming a branching network throughout the body for gaseous exchange and respiration

Index

Bold pagination refers to main references, ***bold italic*** pagination to main subjects of feature boxes and *italic* pagination to captions.

Further Information

Recommended Reading

Atlas of the Dragonflies of Britain and Ireland, R. Merritt, N.W. Moore, and B.C. Eversham. Natural Environment Research Council, HMSO, London, 1996.

Damselflies of North America, M.J. Westfall, and M.L. May. Scientific Publishers, Gainesville and Washington, 1996.

Dragonflies: Behaviour and ecology of Odonata, Philip S. Corbet. Cornell University Press, Ithaca, New York, and Harley Books, Colchester, 1999.

Dragonflies: Naturalists' Handbooks 7, Peter L. Miller. Richmond Publishing Co., Slough, 1995.

Dragonflies of the World, Jill Silsby. Natural History Museum, London and CSIRO, Collingwood, Australia, 2001.

Dragonflies Through the Binoculars: a field guide to dragonflies of North America, Sidney W. Dunkle. Oxford University Press, Oxford, 2000.

Field Guide to the Dragonflies and Damselflies of Great Britain and Ireland, Steve Brooks & Richard Lewington. British Wildlife Publishing, Rotherwick, Hampshire, 3rd edition, 2002.

Hong Kong Dragonflies, K.D.P. Wilson. Urban Council of Hong Kong, 1995.

The Australian Dragonflies, J.A.L. Watson, G. Theischinger and H.M. Abbey. CSIRO, Canberra, 1991.

The Dragonflies of Europe, Richard R. Askew. Harley Books, Colchester, 1988.

The Dragonflies of New Zealand, Richard Rowe. Auckland University Press, Auckland, 1987.

The Dragonflies of North America, J.G. Needham, M.J. Westfall, and M.L. May. IORI, Gainesville, Florida, 2000.

Useful Addresses

Australian Dragonfly Society
Denis Reeves, 30 Bramston Terrace, Herston, Queensland 4006, Australia

British Dragonfly Society (BDS)
Dr W.H. Wain, The Haywain, Holywater Road, Bordon, Hampshire GU15 0AD, UK
http://www.dragonflysoc.org.uk/links.htm
The BDS produces a journal and a newsletter twice a year. It holds many field trips throughout the summer and an annual indoor meeting in November. There are many links to other dragonfly websites on the BDS website.

Dragonfly Society of the Americas (DSA)
Dr T.W. Donnelly, 2091 Partridge Lane, Binghamton, NY 13903, USA
http://www.afn.org/~iori/dsaintro.html
The DSA has an annual meeting and collecting trip, and various regional meetings.

S.I.O. Foundation
Dr Bastiaan Kiauta, Editor of Odonatologica, P.O. Box 256, 7520 AG Bilthoven, The Netherlands
An international dragonfly society that publishes the scientific journal *Odonatologica* and *Notulae Odonatologica*
http://www.afn.org/~iori/siointro.html

Worldwide Dragonfly Association
Jill Silsby, 1 Haydn Avenue, Purley, Surrey CR8 4AG, UK
An international dragonfly society that publishes the *International Journal of Odonatology* and the newsletter *Agrion*.
http://powell.colgate.edu/wda/dragonfly.htm

Picture credits